GEOLOGY OF CONNEMARA

AN INTRODUCTION TO THE PHYSICAL STRUCTURE,
ANCIENT ENVIRONMENTS AND MODERN LANDSCAPES
OF PARTS OF NORTHWEST GALWAY AND SOUTWEST MAYO,
TO ACCOMPANY THE BEDROCK GEOLOGY
1:100,000 SCALE MAP SERIES, SHEET 10, CONNEMARA

J.H. Morris, C.B. Long, B. McConnell and J.B. Archer;

with contributions from
G. Stanley and K. Claringbold (Mineral Resources)
D. Daly (Groundwater Resources)
W. Warren (Quaternary).

Edited by
J.H. Morris and C.V. MacDermot.

Published under the authority of the
Director of the Geological Survey of Ireland.

ACKNOWLEDGEMENTS

This report has been synthesised from a more detailed and technical description of the geology of the Connemara and South Mayo region, published by the Geological Survey of Ireland as a separate handbook, entitled "The Geology of Connemara and South Mayo" by C.B. Long, B. McConnell, J.B. Archer and others. It also draws upon concepts developed in the booklet to accompany Sheet 6 in the Geological Survey of Ireland 1:100,000 scale Bedrock Geology map series, "Geology of North Mayo" (Long *et al.* 1992).
The Geological Survey of Ireland acknowledges the full and friendly cooperation of Professor Bernard Leake in granting permission for the one inch to one mile maps of Connemara and South Mayo compiled by himself and coworkers (John Graham, Paul Ryan, Tony Senior and Geoff Tanner), copyright the University of Glasgow, to be used in the compilation of the 1:100,000 scale map accompanying this report.
The map sheet was produced digitally in Arc/Info format by the Cartography Unit.

ISBN 0 9515006 7 8

Diagrams, design and layout
by Cartography Unit.

Printed in Ireland by
Westprint Ltd, Enniscrone, Co. Sligo

Graphic Reproduction by
Lithographic Plate Plan, Dublin 4

CONTENTS

FIGURES

TABLES

PLATES

AGE (Million Years)	EON	ERA	PERIOD		EVENTS RELATING TO SHEET 10*	OROGENY
1.6	PHANEROZOIC	CENOZOIC	Quaternary		A series of ice ages,followed by spread of vegetation, growth of lowland bogs, arrival of man.	
65	PHANEROZOIC	CENOZOIC	Tertiary	△	Erosion.*Spreading open of north Atlantic from about 65 million years ago, with generation of oceanic crust.* Intrusion of dolerite dykes and plugs.	
135	PHANEROZOIC	MESOZOIC	Cretaceous		Erosion. Probable incursion of 'chalk' sea, with chalk deposited widely *and preserved in N.Ireland. Continuing deposition in offshore basins.*	
205	PHANEROZOIC	MESOZOIC	Jurassic		Erosion and uplift. *Sediments deposited in offshore rift basins and in N. Ireland.*	
250	PHANEROZOIC	MESOZOIC	Triassic		Desert conditions on land, *with sand and gypsum deposited in east of Ireland. Initial rifting (Triassic) between future North America and Europe.*	
290	PHANEROZOIC	UPPER PALAEOZOIC	Permian			
355	PHANEROZOIC	UPPER PALAEOZOIC	Carboniferous	△ ■	Land progressively submerged; coastal sands delta building, limestones deposited in tropical seas, followed by deposition of nearshore then deltaic mud and sands.Late Carboniferous intrusion of dolerite sheets and effects (gentle folds and block faulting) of distant Variscan mountain building (continuing into Permian).	D6 VARISCAN
410	PHANEROZOIC	UPPER PALAEOZOIC	Devonian	▲△ ■	Acadian mountain building episode, rapid erosion, and deposition of 'red beds' in semi-arid conditions.Early Devonian Intrusion of Corvock granite, Galway granite and associated plutons.	D5 ACADIAN & CALEDONIAN
438	PHANEROZOIC	LOWER PALAEOZOIC	Silurian	▲ ■	Sedimentary deposition in shallow remnant sea and its margins, following closure of Iapetus Ocean. Major sinistral shearing and Caledonian mounting building.	
510	PHANEROZOIC	LOWER PALAEOZOIC	Ordovician	▲ △ ▲ △ ▲ ■	Peak regional metamorphism and major deformation of Dalradian schists in early Ordovician during Grampian mountain building. Sedimentation, northward ophiolite obduction, and development of magmatic arc above an initially southward, then northward directed subduction zone as Iapetus began to close. Taconic mountain building. Cessation of subduction with continent-continent collision: effective closure of Iapetus Ocean.	D4 TACONIC D3 D2 GRAMPIAN
544	PHANEROZOIC	LOWER PALAEOZOIC	Cambrian	///	Spreading open of Iapetus Ocean from late Neoproterozoic/early Cambrian with generation of oceanic crust. Initiatian of Grampian (i.e. earliest Taconic) mountain building in late Cambrian.	
1000	NEOPROTEROZOIC			PRECAMBRIAN /// ■	Sedimentary deposition of Dalradian rocks 750-600 million years ago,interrupted by an ice age and periods of volcanism.Detrital zircons 1700-1300 million years old occur in Dalradian sediments. Younger plutons contain 1960-1700 million year old, and Grenville age zircons. Earliest deformation and metamorphism of Dalradian rocks in late Neoproterozoic (Brazilide? mountain building event).	D1 (BRAZILIDE?) GRENVILLIAN
1600	MESOPROTEROZOIC			PRECAMBRIAN	*Grenville mountain building spanning Mesoproterozoic/Neoproterozoic boundary (1100-900 million years ago), reworked the Ketilidian crust. Earliest rifting about 900 million years ago, prior to opening of Iapetus.*	
2500	PALAEOPROTEROZOIC			PRECAMBRIAN	*Igneous material added to the crust during Ketilidian mountain building about 1800-1900 million yearsago,subsequently to become the gneisses of Inishtrahull and the Annagh gneisses of north Mayo. Oldest known rocks and minerals in Ireland about 1800 and 1960 million years old respectively.*	KETILIDIAN
4000	ARCHAEAN			PRECAMBRIAN	*Oldest known rocks on Earth about 4000 million*	
4600	PRISCOAN			PRECAMBRIAN	*Oldest known minerals (zircons) on Earth about 4280 million years old. Formation of Solar System about 4600 million years ago.*	

Table not to scale.

■ Rocks deposited at this time in Sheet 10 area.
/// Deposition, though uncertainty of age.
△ Period of volcanism in area of Sheet 10
▲ Period of intrusive igneous activity in area of Sheet 10
* *Italics refer to areas beyond Sheet 10*

Figure 1. Geological Time Scale

PREFACE

The Connemara and South Mayo region is one of the premier landscape terrains in Ireland, a source of appreciation and enjoyment to countless thousands of visitors every year. But what is that landscape, how did it come into existence, from what is it formed? These are questions to which we hope that this booklet will provide some answers and in so doing guide the reader to a better understanding of the rocky foundation of this part of Ireland: a guide to a heritage written in the rocks, a heritage of very ancient events which have sculpted and moulded the entire face of our planet, including the Connemara and South Mayo region.

This booklet is written to accompany Sheet 10, the new 1:100,000 scale Geological Map of "Connemara" (Note: this title reflects that given to the sheet area by the Ordnance Survey and is used here, even though it includes parts of South Mayo and a small part of North Mayo). Unlike previous Geological Survey of Ireland booklets in this series, the description of the Sheet 10 area has been divided into two distinct publications: this booklet to cater for the non-specialist reader, the other for the specialist reader. The separation has been made, in this instance, in recognition of the diversity and complexity of the geology of the Connemara region which could not otherwise be adequately presented in the booklet format established for the series. Both publications are, however, accompanied by identical versions of the Sheet 10 map.

The area covered by the Connemara map sheet (Sheet 10) has had a long and complex history, extending over about 750 MILLION years. Scientific investigations into such an ancient history, have, by contrast, only been undertaken over the last 180 years, and although much has been accomplished, there are still many unresolved geological problems to be investigated. This account seeks to unravel some of those mysteries and provide an introduction to the evolution of the foundation of the Connemara landscape for non-specialists, as well as provide an introduction to the geology of the region for use by civil engineers and for groundwater applications. The guide is divided into three sections:

Section 1: Reading the geological map of Connemara;

Section 2: The foundation of the Connemara landscape;

Section 3: The earth resources of the region.

Section 1 is, like similar sections in other reports in this series, intended to provide a very brief introduction to the methods used by geologists to represent and depict relationships between rocks on a map. It is not intended to provide an introduction to basic concepts in the Earth Sciences, such as how rocks are classified, how they have been modified and deformed by earth processes or how oceans and continents evolve and change through time. For readers with no previous knowledge of Earth Science, rocks, minerals or fossils, reference is directed to:

- the recent Geological Survey of Ireland publication "Written in Stone" by Dr.P. Kennan (Kennan, 1995). This booklet, written to accompany the 1995 RTE TV Series of the same title, introduces many basic Earth Science concepts in non-technical terms.
- "The Hamlyn Guide to Rocks, Minerals and Fossils" (Hamilton *et al.* 1981). This guide provides considerable introductory information on many different varieties of rocks, minerals and fossils, most of which are illustrated with excellent colour plates.
- or to any of the introductory texts listed in the first section of the bibliography at the end of this booklet.

Section 2, "The foundation of the Connemara landscape" provides a summary account for the non-specialist reader who has an interest in the natural environment and how the landscape of Connemara has been created and how it has evolved. The summary is written in as non-technical a manner as is reasonable, using a minimum of technical jargon. However, some technical terms are unavoidable and where first used, they are highlighted in **bold** and explained in the glossary (Appendix 2). All figure, table and plate numbers are determined from their first reference in this section.

Finally, Section 3 provides a summary of the earth resources of the region, both mineral and groundwater. Even though these sections are slightly more technical, they provide nonetheless an overview both for the general reader, as well as an introduction to the mineral and groundwater resources of the region for more specialist applications.
Two appendices provide extra explanatory information:
Appendix 1: provides a simplified, summary description of all the principal rock units shown on Sheet 10 and is intended principally as an aid for non-specialist users of the map. The terminology follows, as far as possible, that used in GSI Groundwater studies, in order to make information more readily intelligible for such applications;
Appendix 2: provides a glossary of technical terms used, but not explained, in the text.

READING THE GEOLOGICAL MAP OF CONNEMARA

THE ROCK UNITS

Sedimentary rocks are deposited in individual layers called **beds** (or **strata**), which can accumulate to form enormous thicknesses of rock. However, in nature, neither **sedimentary** rocks nor their **metamorphic** equivalents (**meta**-sedimentary rocks) occur in single kinds. Two, or usually more, kinds of sedimentary rock are almost invariably inter-layered with each other: **limestone** with **shale** and/or **chert**; **sandstone** with **conglomerate** and/ or **siltstone** and/or **mudstone**; **psammite** with **meta**-conglomerate and/or **semi-pelite** and/or **pelite**.

To facilitate description and representation on maps, geologists customarily group associated rocks together into a formal hierarchical sequence of units, called **members**, **formations** and **groups**. The formation is the most fundamental of these units, consisting of a sequence of related or similar rock types which can be mapped at the surface or underground. The unit is not scale dependent, although it is normally applied to units of at least several tens of metres in thickness and, in plan view, extending for distances greater than about 1km. Smaller scale, but distinct, rock packages are frequently defined within formations and these are called members. At the opposite end of the scale, related sequences of formations may be combined together into a group, and if really large scale, into **supergroups**. Rocks may also be described, as they are in places herein, by other more informal terms such as **succession** or **complex**, a succession reflecting a rock sequence in a particular location, whereas the term complex is used to highlight an unusual or particularly significant rock association. The study of all aspects of geology relating to the sequence of groups, formations, members and beds is called **stratigraphy**, and of the definition and arrangement of rock units, **lithostratigraphy**.

Lithostratigraphic units are customarily named after some prominent feature of the local or regional landscape. On Sheet 10, for example, the Mweelrea Formation is named after Mweelrea Mountains; the Clew Bay Group after Clew Bay and so on. Formations and members are indicated on the map, and indeed on all others in the series, by a combination of colours and code letters - two capitals for a formation (eg. IR = Kinrovar Schist), with the addition of two lower case letters for a member. The names and position in the stratigraphical succession of the rock units are located in the legend to the map. The name generally includes a geographical term, and sometimes a word naming the main rock type (**lithology**).

Igneous intrusions are distinguished by a different format of code lettering on the map. Their principal lithologies are also shown in the legend.

HOW ROCK UNITS COME TO LIE SIDE BY SIDE

With the passage of time, sediments accumulate, and older strata are buried beneath younger. The present distribution of rock formations is due to earth movements arising from the relative motions of the earth's **lithospheric plates**. If rocks are tilted, then formations deposited on top of each other, like a layer cake, will appear to lie side by side (Fig. 3). But earth movements do more than tilt rocks; they uplift, **fold** and fracture (**fault**) the original "layer-cake" stratigraphy to create a complex mosaic of units. Where folding of an area produces "ridges", such that older rocks appear flanked by younger rocks, this indicates uplift of that region allowing a view of a deeper part of the earth's **crust**: such "ridges" are called **anticlines**. A **syncline** is the opposite, where rocks are folded down into "troughs" so that the younger rocks are flanked by older rocks. Much more complex patterns than just simple faulting or folding reflect plate movements and collisions, and are well exemplified by the very complex deformational, structural and recrystallization patterns evident in the older (Precambrian and Lower Palaeozoic) rocks of Sheet 10.

BOUNDARIES BETWEEN ROCK UNITS

Given the many various ways in which rock sequences can be constructed and then come to be juxtaposed, it is not surprising that there are a number of different types of boundaries generated during the assembly process. These can however, be simplified down into three principal types: stratigraphic, **unconformities** and faults.

Stratigraphic boundaries are, as the name implies, the contacts between different rock units at the time of their formation. They include contacts between beds, members, formations and groups and so on, so long as these have not been modified in any way since they were first formed. Igneous intrusive or **extrusive** contacts fall into the same general category, although intrusive rock contacts frequently cross-cut other types of boundaries, reflecting the manner in which intrusions puncture through other rocks. On the map, all these contacts are depicted by a thin black line and **dip** and **strike** symbols show which way the rocks are inclined.

Unconformities are similar to normal stratigraphic boundaries but, very importantly, they generally reflect a considerable interval between the time of formation of the rocks above and below the contact. They usually represent a large erosional break or gap in the rock sequence, often with an angular discordance between the units due to earth movements which folded the older rocks before the younger ones were deposited.

It is important to realise that unconformities, or gaps in the rock record, represent much the greater part of geological time: such contacts can mark the passage of many tens, hundreds or even thousands of millions of years. Consequently, any history of the earth at a given location on its surface is usually very incomplete - as is clearly the case in Connemara (Fig. 1).

Faults represent fractures which cut through rock sequences and displace the rocks on either side of the fracture - as is abundantly clear every time the San Andreas Fault in California moves. Displacement of rocks on one side of a fault plane, the plane of fracture, is relative to that on the other side and may be up, down, horizontal or, most commonly, oblique. An observer standing on one side of a fault, whose surface dips away from the observer, will in a **normal** fault, notice that the rocks on the far side of the plane have moved down in relation to the near side. In contrast, the opposite sense of movement will be observed in the case of **reverse**, or **thrust** faults (distinguished on the basis of the steepness of the fault plane, high angles for reverse faults, low angles for thrusts): the rocks on the far side of the fault plane will have moved upward towards the observer's side. Normal faults reflect extension in the earth's crust, whereas reverse and thrust faults reflect compression. Lateral displacement faults are called either **wrench** (synonym "strike slip") faults, and oblique sense faults are simply called oblique faults.

The scale of movement on faults is extremely variable ranging from a few metres or so to tens, hundreds, even thousands of kilometres. Faults may reflect either brittle fracturing, akin to fractures in glass or pottery, or, if developed under high pressures and temperatures, they may reflect **ductile** displacement, akin to stretching and displacing plasticine. Faults developed under the latter conditions can show many special features and although the same general terminology is applied to them, special names are applied in some instances, for example ductile faults are frequently called **slides** and ductile wrench faults may be called **shears** or **shear zones**. On the map, faults are generally indicated by thick black lines, although special symbology may also be added to indicate normal, reverse, thrust and wrench faults.

Where metamorphic rocks are found (e.g. in the Dalradian and older rocks), interpretation is far more difficult. The rocks are invariably intensely folded, reflecting multiple episodes of superimposed folding, and the stratigraphy is, consequently, difficult to unravel with any degree of certainty. To add further complications, groups, formations and successions are frequently separated by slides which can juxtapose widely different rock sequences.

CONSTRUCTING A THREE DIMENSIONAL PICTURE FROM THE MAP

By using information on the face of the map, such as the relative ages of formations, the nature of the contacts between them, and the strike and dip (trend and inclination) of bedding, it is possible to construct an interpretative cross-section to indicate the relationship between rock units below the surface. Often, the strike of beds is roughly represented by the trend of formation boundaries, but notable exceptions to this occur where beds are nearly flat lying, in which case the lithological boundaries follow the shape of the topography, and where the topography itself is steep and incised. The direction in which the beds dip may be inferred from the direction in which the formations get younger, these directions commonly being the same. Each segment of the cross section on Sheet 10 has been constructed on the basis of these principles.

WHAT THE MAP DOES AND DOES NOT SHOW

The map shows the distribution of geological units at the earth's surface, or predicted to be present beneath the superficial cover of Quaternary deposits. It is constructed from information recorded at surface outcrops, and from boreholes where available. Features such as continuity of bedding, dip and small-scale folding, allow the geologist to predict the extent and distribution of formations where there are no exposures. Uncertainty grows with increasing distance between outcrops. Thus, where rock outcrops are few and far between, for example in bog areas, the map is, at best, an intelligent guess.

VIEWING THE ROCKS

The map does not show individual rock outcrops. Detailed studies by many geologists provide the information from which the Sheet 10 compilation map has been compiled.

Outcrops or generalized areas of outcrop are shown on many of the large scale original manuscript geological maps surveyed by individual geologists. Some of these, mainly 6 inches to 1 mile (1:10,560) scale maps surveyed in the 19th as well as 20th Centuries, are available for examination in the Geological Survey of Ireland, as are some 1:25,000 scale compilation maps. Others are not readily available, except as plates or diagrams accompanying M.Sc. or Ph.D. theses housed in various universities, most commonly Bristol and Glasgow for Connemara; and Galway (National University of Ireland), London, Dublin (Trinity College) and Oxford, for South Mayo.

All 10 figure grid references (prefixed GR) mentioned in the text are derived from the Ordnance Survey of Ireland 1:126,720 topographic map series. The references therefore define a 10m square. Most of Sheet 10 falls within the Ordnance Survey grid sub-zone L, although a 5km wide strip down the eastern edge falls in sub-zone M.

Symbols on the face of the map showing dip and strike or structural data are placed as near as possible to the locality to which they refer. Those who wish to examine the rocks, will have to visit the good coastal outcrops; search on mountain and hill slopes where outcrops of rock are exposed; in the beds of rivers and streams which have cut through Quaternary deposits; or in quarries or small pits where stone was once worked. Places where typical examples of many units and features may be examined, are given throughout the booklet. The descriptions in the text are generalized, and consequently the lithology (-ies) observable at any point may not be typical of that recorded on the face of the map.

It should be remembered that there is no right of access to examine rock outcrops on private land. Permission to enter upon or cross private land should be obtained from the land owner.

Equipped with the geological map and a little imagination, we can now look at a rock outcrop and see in it something of the passage of time, of the coming and going of mountain ranges and seas of which no evidence remains, except for the record in stone.

THE FOUNDATION OF THE CONNEMARA LANDSCAPE: AN INTRODUCTION TO ITS PHYSICAL EVOLUTION

The area covered by the Connemara map sheet contains some of the most spectacular and varied landscapes in Ireland. Rugged mountains, often with much bare rock, broad glaciated valleys and sheltered mountain glens contrast with large stretches of low bogland, with their abundance of small loughs. A long coastline with many islands presents an almost endless variety of features from high cliffs and deep inlets to swarms of partially drowned **drumlins**.

Each of these distinctive landscapes owes its physical expression to the rocks out of which it is carved and, in turn, to a mosaic of long since vanished ancient environments which gave rise to those rocks. The present and the past link together to form a continuous spectrum of earth heritage and ever evolving and changing environments. This is a history of a small part of Ireland which extends over not just hundreds and thousands of years, but to tens, hundreds and even thousands of millions of years, long, long before the first appearance of humans on planet Earth. We hope that this relatively brief introductory account will reveal a "flavour" of an otherwise all too easily overlooked component of our natural heritage. To do this we have to look at the "history written in stone".

Rocks record the comings and goings of oceans and continents, volcanoes, mountains, ice ages, deserts and "Amazon proportion" rivers. We take for granted the apparent permanence of such co-existing environments in our present day world, a view undoubtedly reflecting the relatively short human life-span. Nothing could be further from the truth. Present day environments, even the configuration and patterns of oceans and continents, are, in the context of earth history, almost as transitory as the human life-span.

This serves to introduce one of the most fundamental, but equally difficult to grasp concepts in earth history: the enormous time scale involved. The oldest currently known rocks on Earth are about 4,000 MILLION years old, the oldest indicated rocks in Ireland about 1,700 MILLION years old and in this region, at most only (!) about 750 MILLION years old. The most recent, "Quaternary", ice age, which sculpted and moulded most of the modern day landscape of Ireland, started about 1.6 MILLION years ago and the last ice sheets which covered most of Ireland melted about 10,000 years ago. In contrast, humans have been "on the scene" in Ireland for only a paltry 9,000 years.

Each particular time interval of Earth History, like normal historical periods, is given a specific name, although some of the finer details may differ from one country or continent to another. Some of these names are likely to be already familiar to the reader, particularly "Jurassic" and "Cretaceous" on account of their connection with Dinosaurs. Others are unlikely to be very familiar, but just as street names are important to help us find our way around a town or city, so too these names are important to help us navigate our way through the history of the Earth. The names used here are indicated in Figure 1, and a little time spent now in familiarisation with the names and time intervals will greatly assist understanding of subsequent sections. Alternatively, just keep referring back to Figure 1.

The second major concept concerns how rocks are categorised and classified. At a gross level, rocks are divided into three major groups, reflecting how they have been formed:

sedimentary rocks - formed by deposition of sediment from water, wind or ice;

igneous rocks - formed by consolidation of molten **magma** derived from deep within the Earth either on (as **lavas**, or extrusive igneous rocks) or beneath (intrusive igneous) the earth's surface; and

metamorphic rocks - rocks derived from either of the other types and transformed, at high

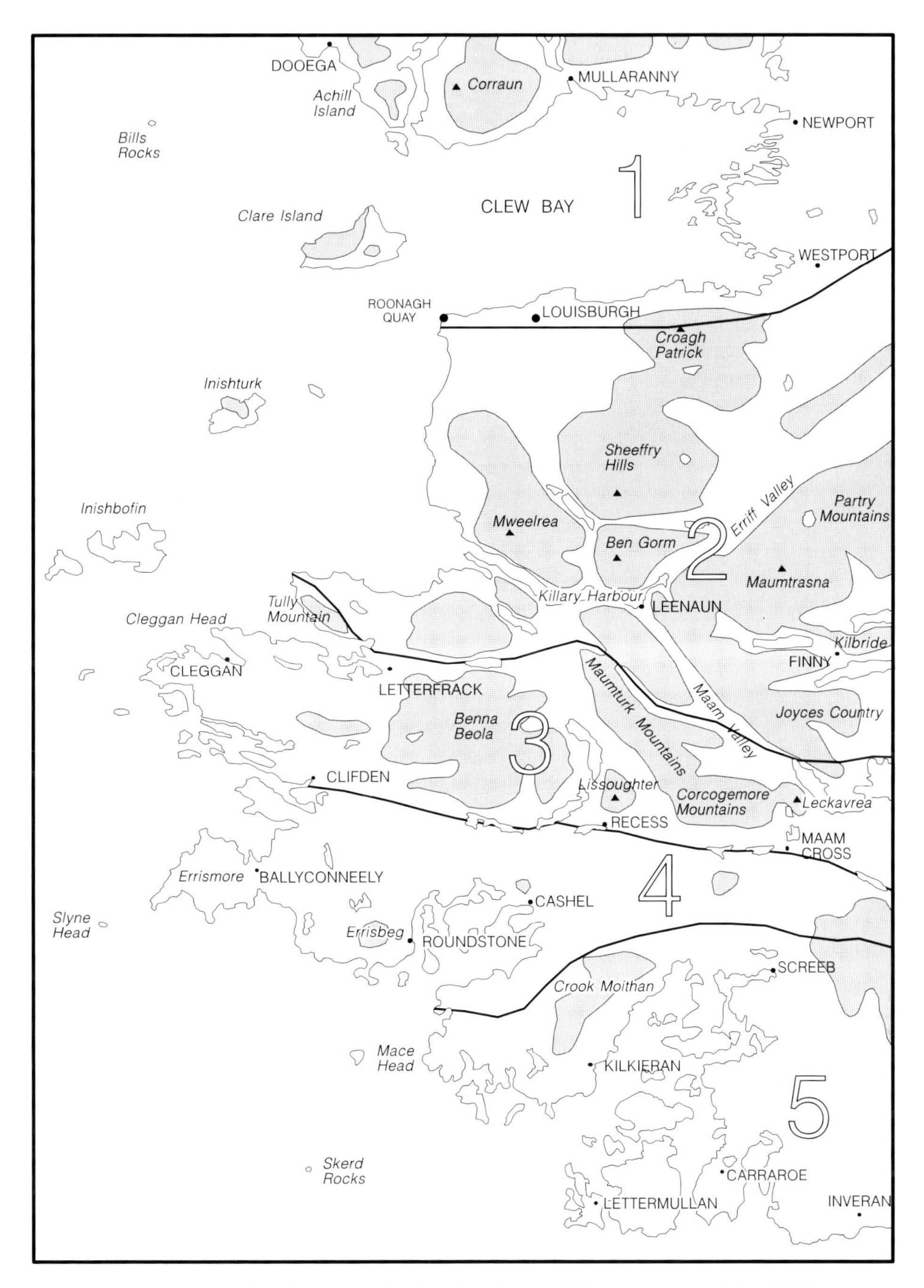

Figure 2. The major landscape and upland regions of Connemara and South Mayo.

temperatures and pressures, but below those needed to melt them, into physically and mineralogically different rocks.

Further consideration of these rocks and how they are classified is referred to the booklet "Written in Stone" and to the general texts listed in the first section of the Bibliography. Any reader inexperienced in basic earth science concepts is strongly referred to these sources of supplementary information before proceeding further in order to appreciate better the following sections.

CONNEMARA: THE FOUNDATION OF THE LANDSCAPE

Solid rocks form a foundation and framework to the landscapes we see now. This might seem like stating the obvious, but it is all to easy to view a landscape and see nothing but shapes draped with a relatively superficial biological and human veneer. But almost invariably, it is the rocks just beneath the ground surface which control or influence the shapes we see and vegetation and human use patterns. The Aran Islands, just to the south of this map area, provide a remarkable example of the relationship between land use, settlement patterns, landforms and the limestone bedrock. And, in its turn, that limestone owes its existence to a particular, ancient and long since vanished geological environment.

Connemara and South Mayo might not possess such striking examples, but even so the area may be subdivided into a number of east - west trending landscape regions, which are all different from one another, and each of which reflects differences in their rocky foundations. From north to south, these are (Fig. 2):

1. The Clew Bay area, guarded by the isolated sentinel of Clare Island to the west, floored by easily eroded limestone and drumlins in the east, and flanked to north and south by mountains constructed from Dalradian metamorphic and Silurian sedimentary rocks respectively.

2. South Mayo with relatively smooth grassy mountains and hills of Ordovician and Silurian sedimentary rocks;

3. Connemara's rugged mountainous spine formed of hard Dalradian metamorphic rocks, including abundant white **quartzite**;

4. Connemara's undulating boggy lowlands, all underlain by other types of metamorphic rocks;

5. Connemara's southern **granite** country with gently rounded, rising ground covered by **blanket bog**.

The hilly and mountainous regions are formed of geologically ancient rocks, although the hills and mountains themselves are of no great geological antiquity. They were gradually uplifted just a few tens of millions of years ago. Since then they have been deeply sculpted by erosion, first under the sub-tropical climate of the Tertiary geological period and then under the rigorous conditions of the last Ice Age, which began about 1.6 million years ago.

Metamorphic rocks underlie a variety of landscapes. For example, ancient, resistant quartzites form the mountainous terrain of the Benna Beola (Front cover plate), Maumturk and Corcogemore Mountains and also the isolated hills at Cleggan Head, Tully Mountain, Leckavrea and Derry, but elsewhere, as for example between Lissoughter Hill and the Corcogemore Mountains, they occupy low ground (Fig. 2). Quartzites and similar rocks also form the Croagh Patrick range south of Clew Bay, the hills of the Corraun Peninsula, and the hills of Joyces Country (Knocknagussy, Kerrygleagh and Ben Levy). Other hard metamorphic rocks underlie the often low boggy ground south of the N59 road from Oughterard to Clifden, and the Errismore promontory near Slyne Head. Softer metamorphic rocks, such as pelitic **schists**, are commonly found in valleys but sometimes form hills, for example in the western Benna Beola and the Knockaunbaun ridge south of the Maam Valley.

Plate 1. The head of Killary Harbour looking northeast along the Erriff Valley. The Erriff Valley fault controls the orientation of the glacially sculpted valley and the upper part of the harbour, which widens significantly where the Maam Valley fault zone intersects the Erriff Valley fault. Note the flat summit level of the tableland on the upper right hand side, southeast of the Erriff Valley.. Photo by J.M. Pulsford, 1982.

Plate 2. Corries on the southeast slopes of the Mweelrea Mountains, above Bundorragha. Tors of Ordovician Rosroe Formation conglomerates in the foreground. Photo by J.M. Pulsford, 1982.

The igneous rocks, despite their hardness, generally do not form conspicuous features in the landscape, although there are some exceptions: the Corvock Granite of South Mayo; several relatively small hills in Connemara, such as Benchoona, Currywongaun, Doughruagh, Doon, Errisbeg and Cashel; the granite forming the high ground north of Kilkieran (Crook Moithan area), and the **volcanic** rocks of the Kilbride peninsula (Knock Kilbride) and Bencorragh.

In South Mayo, Ordovician and Silurian sedimentary rocks have been uplifted and shaped to form high ground (e.g. Mweelrea, Sheeffry Hills and Partry Mountains - Fig. 2)), but elsewhere they usually occupy low ground. When newly uplifted, the South Mayo uplands evidently formed a tableland which extended from Lough Mask westwards to the Atlantic coast and from Clew Bay southwards beyond Killary Harbour. The original plateau surface still dominates the skyline over Lough Mask and Killary Harbour (Plate 1) and elsewhere survives atop the mountains.

Dissection of the plateau and the formation of the modern mountainous landscape was finally accomplished by glacial erosion. Frost-shattering and ice-scouring excavated great linear depressions (Plate 1) along lines of structural weakness caused by faulting (Maam Valley, Doo Lough Valley, Erriff Valley and Inagh Valley - Fig. 2); where softer rocks provided easy prey for erosion (Glenummera Valley), and where faulting and weak rocks combined to facilitate erosion (Killary Harbour and Killary Bay Little). High on the mountain slopes huge, scoop-shaped **corries** were gouged out by glacial erosion and appear like giant thumb-prints on the uplands (Plate 2).

While glacial erosion gnawed at the upper slopes, deposition occurred on lower ground. Quaternary sediments form a discontinuous but often thick blanket concealing much of the **bedrock** throughout the map sheet. Glacial meltwater sands and gravels are common east of Letterfrack, south of Leenaun and flanking the Erriff valley, and drumlin fields are conspicuous at the eastern end of Clew Bay. Evidence of the sculpting effect of the ice can be seen in corries, and in moulded or smoothed rock outcrops known as **roches moutonnées** (literally "sheep rocks", after the form of reposing sheep). These sediments and landforms provide the immediate substrate for the eventual growth of the low-level blanket bog which typifies so much of the landscape of South Mayo, and for the development of the soils which support the oases of green fertile fields set amidst the otherwise barren landscape of the West of Ireland.

Areas underlain by different types of Devonian and Carboniferous rocks may also be readily distinguished by differences in land fertility. Areas underlain by Devonian or Carboniferous sandstone or conglomerate are generally covered by boggy, infertile land, whereas more fertile landscapes are generally underlain by younger Carboniferous limestones.

ENVIRONMENT AND LANDSCAPE CHANGE

The landscapes we view today are the sum of many changes, some of them very recent. Perhaps the most striking example of recent change is the afforestation of vast areas, thereby concealing underlying contours, solid rocks, and superficial sediments. Looking beyond our own very recent modifications to the landscape, we cannot fail to marvel at the changes brought about by glaciation during the ice age.

In Ireland, we are inclined to assume that the rocks are a permanent, unchanging backbone to the landscape as, in our lifetime, little may happen to change our view. If, however, we lived in a currently geologically more active part of the world, such as the Pacific coast of the United States of America or Japan, we would have a different outlook; earthquakes, volcanic eruptions, and the consequent changes to the landscape are expected in such a region. Rapid changes have taken place, however, over the area of Sheet 10 in the pre-historic past.

Neolithic settlements discovered beneath 2m thick bog at Keadue in North Mayo, indicate significant changes to the landscape only some 4,500 years ago. A climatic deterioration brought about wetter conditions affording ideal conditions for the growth of bog following a succession of wet summers. The North Mayo bogs also provide evidence of distant volcanic eruptions in the form of c. 4,000 year old ash layers, the precise source of which has yet to be defined (V. Hall *personal communication* 1995).

Major geological changes, however, generally take place over huge time spans measured in millions of years. These major changes involve the movement of the earth's lithospheric plates over the earth's surface, sometimes accompanied by their fragmentation and mutual collisions (Appendix 1). The geological history of "Connemara" spans about 750 million years, but certain geological events can be traced back to about 2000 million years ago by studying the more subtle clues preserved within the older rocks.

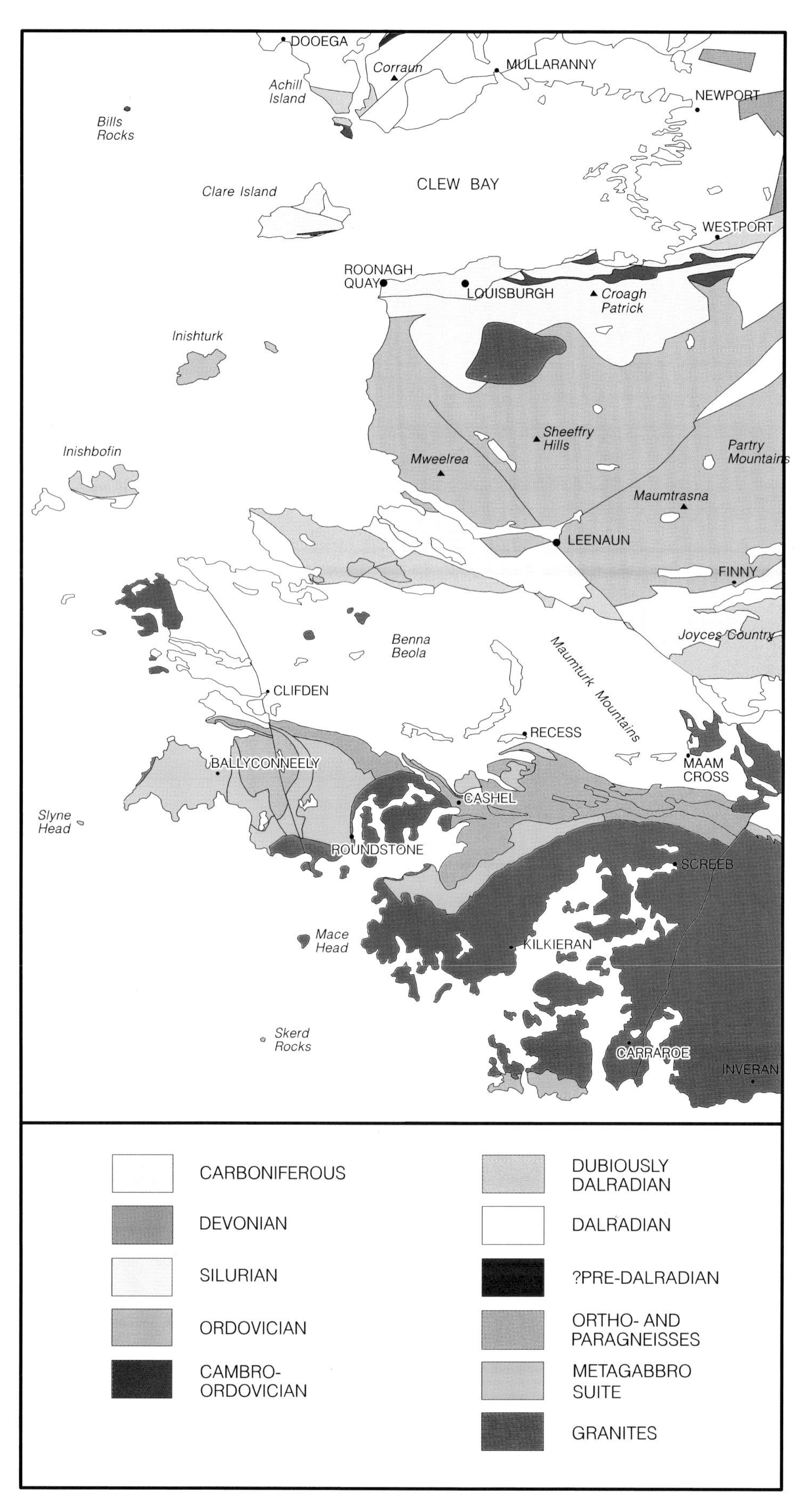

Figure 3. The generalised geology of the Sheet 10 area.

GEOLOGICAL AGE AND THE CONNEMARA MAP

In a broad sense, the ages of the rocks represented in each of the landscape regions noted earlier are broadly divisible into several distinct major categories (Figs 2 and 3). These are:

- the metamorphic rocks of Late Precambrian ages, mainly Dalradian, which date from c. 750 to c. 600 million years ago (dubiously Dalradian, Dalradian and ?pre-Dalradian areas on Fig. 3);

- metamorphic rocks, possibly of Cambro-Ordovician age, c. 550 to c. 440 million years old;

- the mildly metamorphosed sedimentary rocks of Lower Palaeozoic (Ordovician and Silurian, Fig. 3) age, deposited at intervals between about 510 to 420 million years old;

- sedimentary rocks of Upper Palaeozoic (Devonian and Carboniferous, Fig. 3) age, deposited during the period from 410 to 330 million years ago;

- igneous rocks intruded episodically at four principal times from about 650 to 55 million years ago. The older igneous rocks are also metamorphosed;

- and, finally, relatively recent age Quaternary sediments which obscure many of the rocks shown on the map sheet. These sediments are mainly of glacial origin, and range in age from about 1.6 million years old to the present day.

We can now look at each of these major categories in a little more detail and thereby develop an appreciation of the long extinct environments and land and seascapes which once characterised Connemara. On Sheet 10, by far the most important time interval is represented by the rocks of the DALRADIAN, formed between about c. 750 and c. 600 million years ago. We will, in light of the importance of these rocks to the foundation of Connemara, focus upon them principally and then progress upward through geological time to the youngest. Reference should be made to Figure 1 throughout as a "navigational" aid!

Pre-Dalradian ?: older than 750 million years ago

Possibly the oldest rocks in the sheet area are the pelitic and semi-pelitic schists of the Kinrovar Schist (Fig. 5; IR on Sheet 10). These rocks originated mainly as sediments and their subsequent history of folding and metamorphism is at least as complex as that for the Dalradian rocks. Their exact age is still unknown.

The Dalradian: c. 750 - c. 600 million years ago

The bulk of the metamorphic rocks of the Connemara region are part of a belt of metamorphic rocks which extends from the west of Ireland and Donegal, through Tyrone and northeast Antrim, into Scotland where the rocks underlie the Grampian Highlands. These are the rocks of the DALRADIAN SUPERGROUP which have been the subject of intensive study for nearly two hundred years, principally in Scotland, from where the name derives. The supergroup was named, in 1891, after Dál Riada, the territory of an ancient tribe, which extended from western Scotland to Ulster. Appropriately enough, Dalradian rocks are well exposed in Ulster, particularly in Donegal, and other counterparts have been recognized and defined in Mayo and Connemara. Throughout this large geographical area, the Dalradian is composed entirely of metamorphic rocks, mainly metamorphosed marine sedimentary rocks, with subordinate volcanic and intrusive rocks and even metamorphosed glacial-marine deposits.

For well over a century the age of these rocks has puzzled geologists, largely because they contain almost no fossils, which are the principal time-pieces used by geologists to determine the relative age of rocks. In the absence of fossils, geologists can turn to the relationships with neighbouring rocks of known age to derive relative ages and, in recent times, to **radiometric** age dating which can provide precise ages. There are now sufficient age determinations by such methods, on various components of the Dalradian, for us to be fairly certain about the general timing of successive stages in its evolution.

The story of the Dalradian starts with continental tension and crustal thinning about 750 million years ago. This marked the beginning of fragmentation of what until then had been a single, Neoproterozoic age **supercontinent**, called "Rodinia" (Figs 1 and 4), and the formation of an ocean basin in which the Dalradian sediments accumulated. The continental shelf was initially stable and deposition occurred in shallow seas. As time passed, the shelf

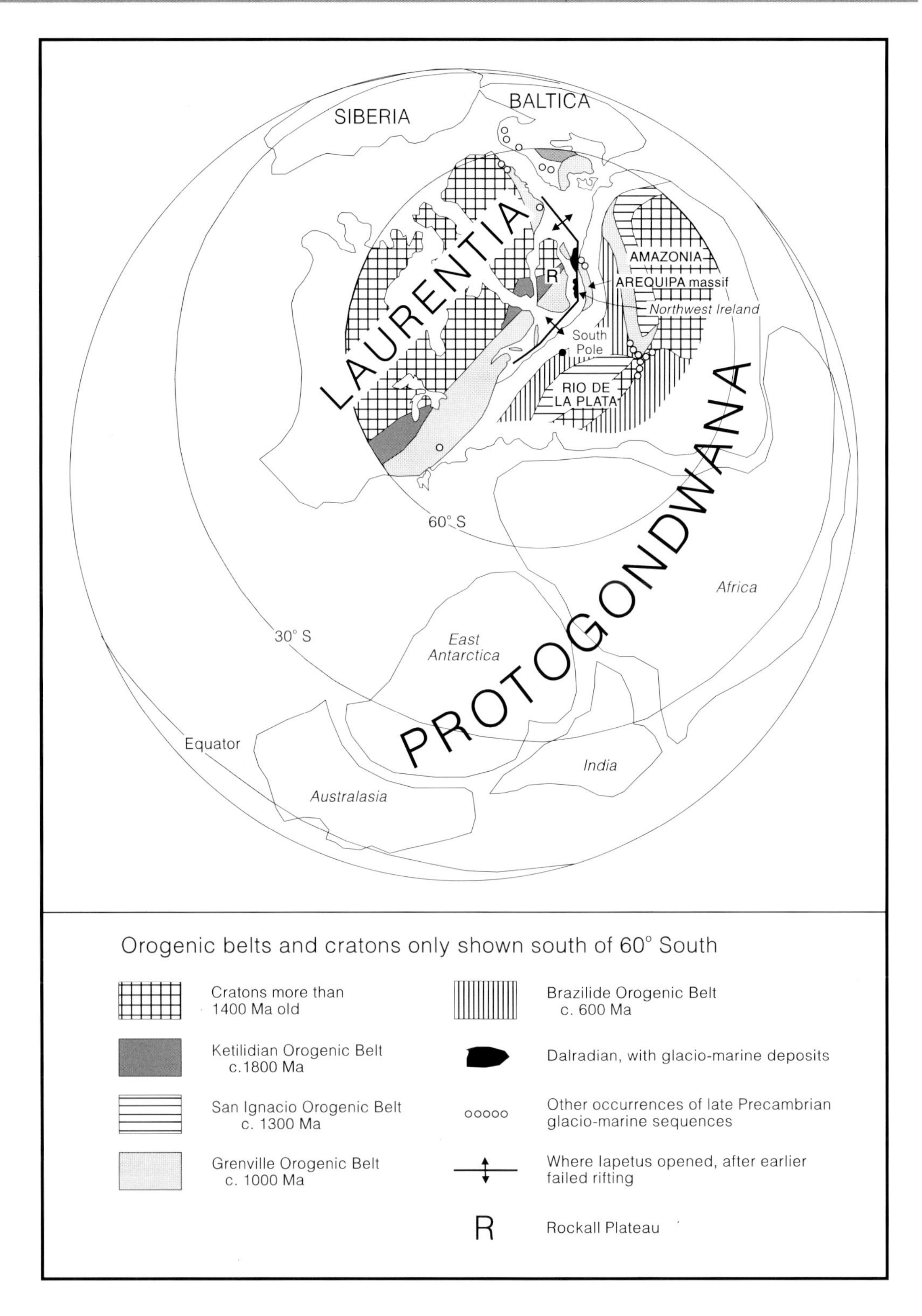

Figure 4. Late Neoproterozoic plate tectonic reconstruction for c. 580 million years ago showing the ancient supercontinent "Rodinia" at the time of its fragmentation (based on Dalziel 1995). Note that northwest Ireland (area exaggerated for clarity) lay between the ancestral North American continent "Laurentia" and what is now western South America, close to the then South Pole, and adjacent to the line along which the supercontinent rifted apart to form a major ocean, "Iapetus". The mosaic of patterned areas depict crustal areas which had been deformed by continental collision and other events during their amalgamation to create "Rodinia" - a cumulative process which took more than 1,200 million years to complete, between 1,800 and 600 million years ago. Names in italics indicate the location of modern day continental landmasses 580 million years ago.

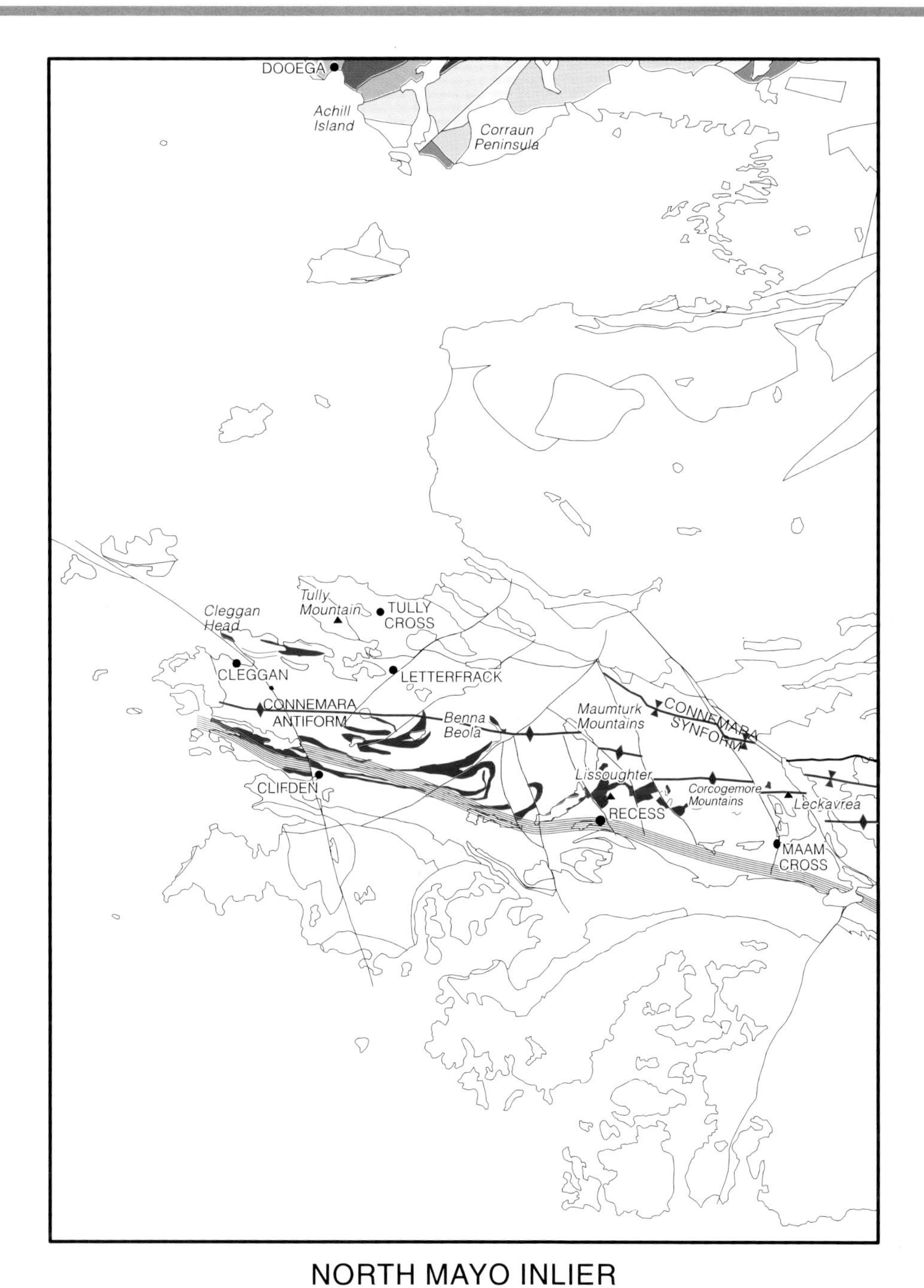

NORTH MAYO INLIER

EAST ACHILL - NORTH CORRAUN SUCCESSION	SOUTH CORRAUN - WEST NEPHIN BEG SUCCESSION	INISHKEA DIVISION
Argyll Group	Appin Group	Kinrovar Schist
Appin Group	Grampian Group	
Grampian Group	EAST NEPHIN BEG SUCCESSION	
	Argyll Group	

CONNEMARA INLIER

CONNEMARA INLIER	BENNABEOLA SUCCESSION
Argyll Group	Steep Belt
Appin Group	

Figure 5. Distribution of Dalradian rocks in the Sheet 10 area.

became more stretched and unstable, and volcanism was initiated as an accompaniment to rifting. This eventually culminated about 600 million years ago with final rupture of the continent and the spreading open of a major new ocean, IAPETUS, which gradually separated two major, ancient continental blocks "Gondwana" and "Laurentia" (Figs 4 and 8).

At the initial stage, block faulting on the rifting (= pulling apart) continental margin created a number of separate depositional basins. Water depths fluctuated, but generally increased, and sediments were deposited on deeper shelf and even on the continental slope. Even so, it is unlikely that the developing ocean was very wide at this stage or that it was floored by true ocean crust.

Here we will describe the rocks of the Dalradian as they are currently exposed in the sheet area (Fig. 5). There then follows an interpretation of what the rocks were and, amplifying the general geography of the time outlined above, the environment in which they were formed. Finally, we relate the events that transformed them to their present state.

SOUTHERN HIGHLAND GROUP
METAMORPHOSED **WACKES**, MUDSTONES AND **BASALTIC** LAVAS: Psammitic, semi-pelitic and pelitic **wackes**, and **basic** metavolcanics.
ARGYLL GROUP
METAMORPHOSED GLACIAL DEPOSITS, QUARTZ SANDSTONES, WACKES, MUDSTONES, LIMESTONES AND BASALTIC LAVAS: basal glacio-marine sequence with **tillites**, overlain by whitish quartzite, followed by interbedded psammitic and semi- pelitic schists, basic metavolcanics, **marbles**, and psammitic and semi-pelitic wackes with aluminous (aluminium rich) pelitic schists.
APPIN GROUP
METAMORPHOSED SANDSTONES, MUDSTONES AND LIMESTONES: Predominance of psammitic schists and quartzites, with pelitic and semi-pelitic schists,**carbonaceous** pelitic schists, and commonly **dolomitic** marbles
GRAMPIAN GROUP
METAMORPHOSED GREYWACKE, **ARKOSIC** SANDSTONES AND MUDSTONES: Greywackes in lower part; upper part more arkosic. Mostly pale coloured quartzites and psammitic schists, with subordinate pelitic and semi-pelitic schists.

Table 1. Summary of principal rock types in the Dalradian Groups. For explanation of rock types, refer to the glossary, Appendix 3.

THE DALRADIAN SUCCESSION

Metamorphosed, generally marine sedimentary rocks comprise the Dalradian Supergroup which is divided into four groups, from the oldest, Grampian Group at the base, to the youngest, Southern Highland Group at the top. The rock types in the groups are summarized in Table 1 and their distribution illustrated in Figure 5. Note that the occurrence of the groups is repeated in different parts of the map area in what are termed "successions".

Grampian Group rocks originated during an early rifting episode about 750 million years ago, which was related to the later opening of the Iapetus Ocean. Appin Group rocks were deposited on a stable continental shelf, as were the very oldest rocks of the Argyll Group. The rest of the Argyll Group and younger Southern Highlands Group rocks were associated with a later phase of rifting, and with the spreading open of the Iapetus Ocean (p.12) prior to the end of Dalradian deposition about 600 million years ago.

The Dalradian rocks crop out in successions separated from one another by faults or major tectonic slides. Their principal area on Sheet 10 is in Connemara. Most of their original sedimentary features have been destroyed, and in some areas they are locally upside-down. In latest Cambrian to early Ordovician times they were deformed into a major mountain belt which subsequently continued to develop throughout the Ordovician period. Metamorphism has been substantial and complex, and repeated folding is common and has given rise to complex structures (Main map, cross sections, and Fig. 5). The rocks are all **schistose** to a varying degree.

The four major Dalradian groups were originally defined in Scotland, and the terminology applied in Ireland. The oldest rocks of the succession, those of the Grampian Group, lie adjacent to the Annagh and Inishkea Divisions in the northwest and the west (north of Sheet 10). These were the first sediments deposited in the deepening Dalradian basin, and are best seen on the Mullet peninsula of North Mayo.

DALRADIAN ROCK TYPES

Quartzites and psammitic schists

White to pale cream to very pale brown, yellow and green quartzites and psammitic schists (composed mainly of **quartz**, with some **feldspar** and **mica**) are the most common rock types in the Dalradian, particularly in the Grampian and Appin Groups.

Both are metamorphosed sandstones which have resisted erosion in most places. They now form the present day chain of mountains which stretches from the west coast of Achill Island, through the Corraun Peninsula to, and including, the Nephin Beg range. Much of the high ground in Connemara (Benna Beola, Maumturk and Corcogemore Mountains, Tully Mountain, Leckavrea, Lissoughter Hill and Cleggan Head) is formed of whitish quartzite, which can be easily examined in the Maam Gap north of Maam Cross (Fig. 2).

Pelitic Schists

Secondary to the meta-sandstones are pelitic schists (former mudstones, now mica-rich and often shiny) and semi-pelitic schists (containing quartz and mica in about equal proportions). Pelites form some of the peaks in the Benna Beola, (e.g. Muckanaght and Bealnascalpa, and also the elongate Knockanbaun ridge between the Maam valley and the Bealnabrack River). This is unusual since pelitic rocks are relatively soft and tend to wear down easily.

Other rock types

Psammitic and semi-pelitic wackes (originally) impure feldspathic greywackes and pebbly wackes) are found in the Argyll Group and represented by the Ballynakill Schist Formation (BK) in Connemara. Marbles (metamorphosed limestones) are found in the Appin and Argyll Groups. White to very pale grey **calcitic** marble can be seen in the Lakes Marble Formation (LM) and locally at the top of the Barnanoraun Schist Formation (BZ). Pale brown dolomitic marble occurs locally within the Connemara Marble Formation and in the low cliffs in Ashleam Bay just south of the path to the beach (Ashleam Bridge Dolomitic Formation, AD). Green marbles are well displayed in the disused Streamstown quarry in the Connemara Marble Formation (CJ), though weathering has now discoloured the once clean and conspicuously coloured walls of the quarry (Plate on back cover of booklet). Within many of the marbles resistant ribs of **calc-silicate** schist can be seen. Basic metavolcanics are found in the Argyll Group (e.g. upper part of the Lakes Marble Formation).

An important **boulder bed** sequence is found at the base of the Argyll Group, and this has been interpreted as a glacial-marine sequence with tillite, resembling a glacial **boulder clay** (Plate 3). This can be correlated throughout the Dalradian of Ireland and Scotland and, on Sheet 10, it is well exposed at Cleggan Head. Where actual tillite is not exposed it can be difficult to assign the rocks to their correct formation or group. Carbonaceous pelites and semi-pelites (e.g. in cliffs at Ashleam Bay on Achill Island, and in the Cornamona area), and calcareous semi-pelites and psammites (e.g. upper part of the Streamstown Schist Formation, ST) are also to be found.

Plate 3. A glacially formed boulder bed, consisting of boulders in a mudstone matrix. Dalradian, Cleggan Boulder Bed Formation, Illion East Townland, western Corcogemore Mountains. Photo by C.B. Long, 1992.

GEOLOGICAL HISTORY OF THE DALRADIAN ROCKS

Crustal Stretching

The very earliest rifting evidence (from Scotland and the Appalachians) is pre-Dalradian and dates back to about 900 million years ago, though true spreading was initiated later. About 750 million years ago, the continental crust under what was to become the Dalradian parts of northwest Ireland and the Grampian Highlands of Scotland began to stretch and thin. The stretching was in response to massive upwelling and sideways spreading of molten **magma** at a **hot-spot** within the underlying **mantle**.

At the surface this stretching was expressed as a slow sinking of the land below sea level. Roughly 700-600 million year old **dykes** in the Annagh and

Inishkea Divisions and the Dalradian rocks of north Mayo (Sheet 6) may well represent magma rising from the mantle at that time. Tensional fissures produced in the crust by stretching would have allowed some of the magma to penetrate toward the surface before cooling as **dolerite**, later to become metamorphosed to metadolerite, and some even reached the surface as volcanic lavas and **tuffs**. The earliest minor basic volcanics occurring on Sheet 10 possibly date from latest Appin Group times in Connemara and slightly later in North Mayo (early Argyll Group), though these could be minor intrusive **sills** rather than lavas. In Scotland, even earlier minor volcanism occurred locally in early and later Grampian Group times and was associated with the early failed rifting event. Further local minor volcanism also occurred in subsequent Appin Group time.

A new Ocean Basin

The sinking of this part of the late Neoproterozoic (Figs 1 and 4) continental landmass led eventually to the formation of a broad ocean. The ocean was of unknown width but extended at least from the Shetland Isles, off northern Scotland, to the west coast of Ireland. It may have ranged as far as Greenland, Norway and the United States of America, and possibly into western parts of South America (e.g. southern Peru and eastern Bolivia). Remember that this was long before the Atlantic Ocean opened up to separate the present continents of Europe and North America, and it was also prior to the opening of the Iapetus Ocean. The present day configuration of the basin is likely to represent a substantial modification of its original shape owing to subsequent plate movements and interactions, including folding and shearing.

The earliest Dalradian sediments were deposited in shallow water (lower Grampian Group). None of these rocks are, for certain, exposed in Ireland. Mid-Grampian Group turbiditic greywackes of the Inishkea Division are partly represented on Sheet 6 to the north of Sheet 10 and reflect basin deepening which accompanied the early, failed rifting. However, unlike other mid-Grampian Group greywackes, they contain a high proportion of sediment derived from igneous source rocks. Late Grampian Group times brought a return to shallow water deposition with arkosic sandstones, as now exemplified by the psammitic schists of the Anaffrin (AN) and Sraheens Lough (SR) Formations.

Estuaries and shallow seas

Despite metamorphic alteration of the original sedimentary rocks, it is still possible to recognize the conditions under which they were once deposited. Erosion of the land beside the new sea produced huge volumes of sediment which was carried by major river systems into the marine basin. **Ripple marks** preserved in the rocks reflect the movement of water currents and suggest that these early Dalradian sediments were deposited in river estuaries along the margin of the basin and on a shallow marine shelf offshore. The activity of currents on the sea-floor winnowed the sands clean of fine muddy material and these sands were later transformed into what we now see as quartzites.

Continued sinking of the basin floor permitted the gradual accumulation of a succession of sediments, mostly sand and mud, several thousand metres thick. Earliest Appin Group times are reflected by shallow water deposits, as exemplified by the quartzites of the lower part of the Cullydoo Formation (CS) and the lower part of the Portnahally Formation (PO). **Desiccation cracks** and ripple marks are present in the latter unit and both contain **cross-stratified** sandstones. Stable shallow water shelf conditions persisted throughout the rest of Appin Group times and into earliest Argyll Group times.

Tropical seas and ice-sheets

Knoll-like structures (**stromatolites**), produced by the activity of **algae** on the sea floor, are preserved in a marble at the very top of the Appin Group. Unfortunately, marbles of equivalent age on south Achill Island (Derreen Marble Formation, DE) and in Connemara (Connemara Marble Formation, CJ) do not preserve such structures. Similar structures found in shallow water, **carbonate** sediments of the present day Caribbean suggest that North Mayo experienced a warm climate during Dalradian times. This particular marble, however, is fairly closely succeeded by a boulder bed sequence containing pebbles, cobbles and boulders (size terms, Appendix 1) of dolomitic marble and granite.

The boulder bed formation is well exposed locally in Connemara, especially at Cleggan Head (Fig. 2; also Plate 3). It is interpreted as tidally and glacially influenced marine sediments deposited below floating sea-ice. The tillite was originally rather like boulder clay, lacking internal bedding, and in some ways similar to boulder clays deposited during the last (Quaternary) ice age, now extensively exposed in such Counties as Cavan and Monaghan.

Some evidence from the Dalradian boulder bed suggests that the source area may have lain to the south.

The juxtaposition of rocks thought to represent tropical and glacial conditions respectively is enigmatic, but it has been thought that widespread late Precambrian glaciation reached low (i.e. equatorial) latitudes and may have been virtually global. A more recent interpretation suggests that Dalradian glaciation occurred in high southern latitudes, therefore apparently indicating a relatively swift climatic change.

Earthquakes and submarine turbidity currents
The continuation of shallow marine conditions after the glacial interlude is suggested by a quartzite (e.g. the Bennabeola Quartzite Formation, BX) which immediately succeeds the tillite. Sedimentation was interrupted sporadically by the activity of volcanoes in an environment somewhat similar to that of the northern part of the present day Red Sea and Gulf of Suez. Soon after, however, stretching of the continental crust reached a climax. Sudden and even greater deepening of the Dalradian basin resulted, accompanied by earthquakes along the faults which controlled this collapse. Sand and gravel carried into the sea by rivers were no longer winnowed clean on a shallow marine shelf. Instead, triggered by earthquakes, the unconsolidated sediments were periodically re-mobilised from the shelf to form submarine **turbidity currents**, high density mixtures of mud, sand, and water which flowed at speed down the continental shelf and slope and out onto the deep ocean floor. There they settled to form **turbidites**, very characteristic types of rock consisting of interlayered impure, mud-rich gravels, sandstones (pebbly quartz wackes and greywackes) and mudstones, which accumulated by slow settling of mud onto the ocean floor between the passing of episodic turbidity currents. Successive turbidites slowly accumulated to form a "submarine fan" (resembling a river **delta**). After metamorphism these became the psammitic, semi-pelitic and pelitic wackes which characterize the Argyll Group, especially its upper part (e.g. Ballynakill Schist Formation, BK).

Volcanic lavas
Sudden deepening of the Dalradian basin was accompanied by rifting and the onset of volcanic activity. This activity reached a peak during mid Argyll Group (ie. Lakes Marble Formation, LM) and later again in end Argyll Group and early Southern Highland Group times, though this later episode is not represented on Sheet 10. Some of the basaltic lavas which poured out onto the sea bed formed **pillow** like masses (Plate 5). Unfortunately these pillow shapes have been destroyed over the area of Sheet 10 by successive periods of deformation.

Igneous intrusions
Dolerite and **gabbro** sills, likely to have been associated with Dalradian volcanism, and now represented by metadolerite, metagabbro or their higher metamorphic grade equivalent, amphibolite, are present in the Dalradian areas.

DEFORMATION OF THE DALRADIAN ROCKS
The Dalradian rocks are highly deformed. A subtly preserved and poorly understood initial deformation during the late Precambrian about 600 million years ago was followed by a lengthy period of the spreading open of the Iapetus Ocean. Then, in latest Cambrian to very earliest Ordovician times, about 510 million years ago, the plates on either side of the Dalradian basin started to converge. The Dalradian rocks were compressed between the advancing basin margins in several stages over the next 100 million years as Iapetus closed again (see section, p.27).

Each phase of convergence (or **orogeny**) crumpled, folded and refolded the layered rocks so that they were repeatedly thickened and pushed up to form a huge mountain chain, much as the Himalayas are being formed today by the comparatively recent convergence of India with Tibet. At the same time the base of the sedimentary pile was pushed down to great depths within the crust. Enormous temperatures and pressures produced by the convergence brought about the metamorphic recrystallization of the original sediments into schists. One or more generations of folds were formed during each convergence, and each phase of folds refolded those that had gone before it, so that complex patterns were developed in the schists (Fig. 6, Plate 4). Fold axes are not distinguished by age on the map. All fold axes have been omitted from the Dalradian area of Connemara for clarity. At several different times quartz **vein**s were formed from **silica** released from the rocks as a result of the metamorphism and deformation, and various igneous rocks were intruded as the plates converged. A similar sequence of events can be recognized in both North Mayo and Connemara Dalradian rocks.

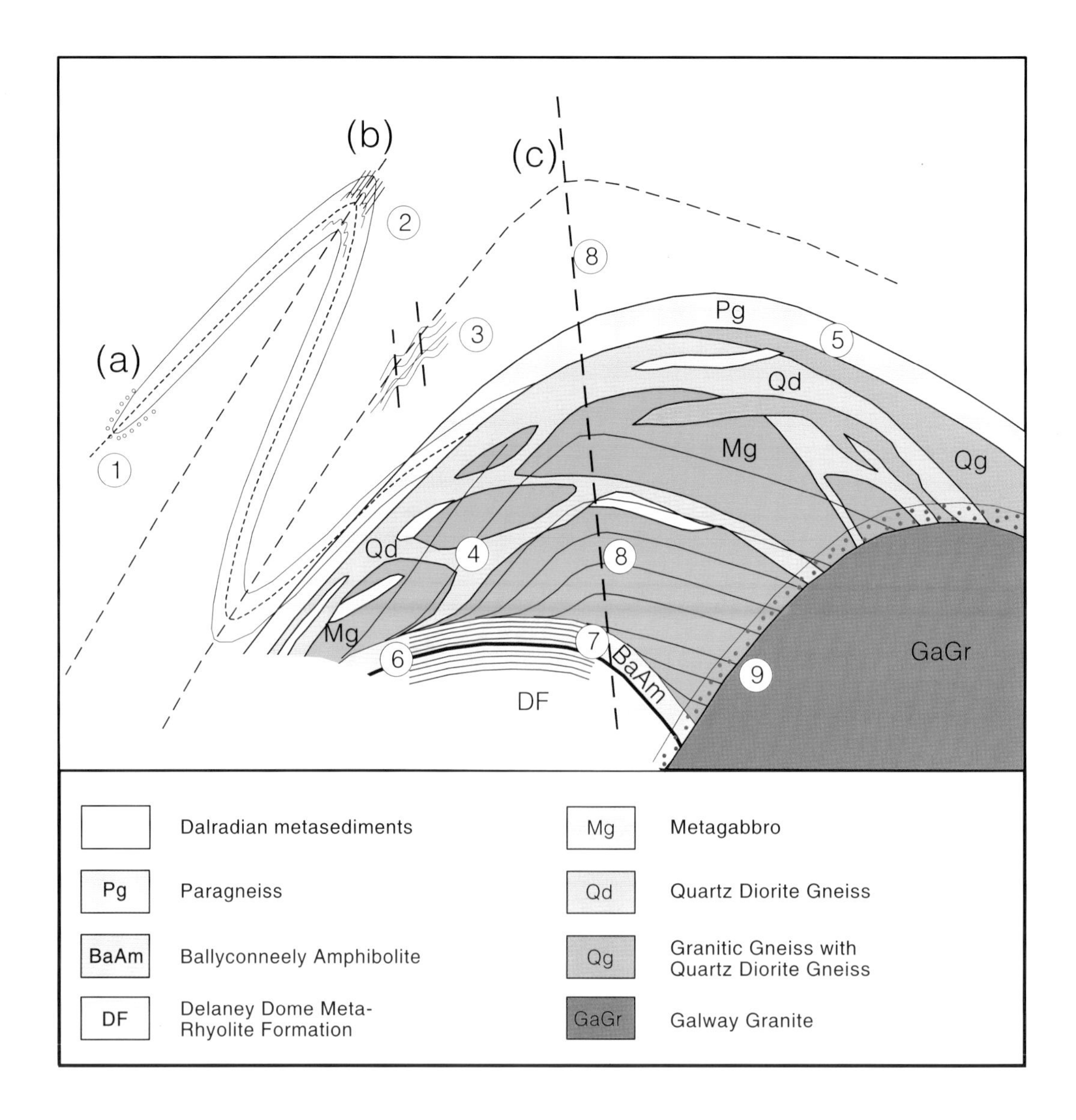

Figure 6. Diagrammatic representation of selected, successive (1 to 9) structural relationships which reflect, in cartoon style, the major events involved in the physical construction of Connemara. Note how the Dalradian rocks are deformed by early, very narrow, parallel-sided folds (a,1) which have been refolded (b,2). The rock units marked Mg, Qd and Qg all represent different types of basaltic to granitic igneous intrusive bodies emplaced into the Dalradian rocks deep in the bowels of an early Ordovician volcanic arc - the "**magmatic arc**" described below. These intrusions thoroughly metamorphosed the meta-sediments into which they were intruded, to create the type of rocks marked Pg (5), and all were then buckled and warped by further folds and related structures (3, 4) This entire sequence, the Dalradian of Connemara and the volcanic arc within it, were then thrust en masse over other remnants of the early Ordovician volcanic arc (DF) along a fault called the Mannin Thrust (6). The thrust is a zone of very intense deformation, marked by a very strong planar layering in rocks above and below the fault and by inversion of rock just above it (7). The complete rock assemblage was then folded yet again (8 or c) and finally (!) intruded by the Galway Granite (9) which locally metamorphosed adjoining rocks of all types in a narrow **aureole** (red stipple). No scale is implied, and relative size has no significance.

Plate 4. Early parallel sided fold, refolded by a later, tight fold in the Dalradian, Ballynakill Schist Formation, Letterbreckaun Townland, north of Lough Inagh. Compare with Figure 6, folds (a) and (b). Lens cap diameter 5.5cm. Photo by C.B. Long, 1992.

Dubiously Dalradian (possibly Cambro-Ordovician): ? 550 - 470 million years ago

The geologically complex rocks which form the mountainous areas of South Mayo include several rock sequences which are imprecisely dated and/or subject to ongoing debate as to their origin and affinity. The sequences may, for convenience be grouped into two distinct suites of rocks, one exposed in south Achill, north Achillbeg, near Westport and in the area stretching from Joyces Country through North Connemara to Inishbofin; the other on Clare Island and along the south shore of Clew Bay from just east of Old Head in the west to near Westport in the east.

The first suite of rocks resembles the Dalradian in many respects and, at least in part, have commonly been considered part of that sequence: in Section 2 and on the map, they have here been termed "dubiously Dalradian" (Fig. 7).

The other suite, although very disparate both geologically and geographically, has always been distinguished from the Dalradian and generally inferred to be of Cambro-Ordovician age. On the geological map, and in Section 2, the presumed Cambro-Ordovician rocks are grouped together under the heading "Cambro-Ordovician" and are described separately in the next section.

The "dubiously Dalradian" rocks have mostly all been metamorphosed and deformed to levels and styles similar to that of the Dalradian. Metamorphic equivalents of turbiditic sandstones, siltstones and mudstones, locally with volcanic rocks, predominate in some sequences and may include minor rock types such as **serpentinite**, **mélange** and marbles. In Joyces Country (Fig. 7), the sequence is dominated by the metamorphic equivalents of sandstones, siltstones and mudstones, carbonaceous mudstones, and other metamorphic rocks such as quartzite and marbles, some of which contain possible fossil remains of algae. The lower part of the latter sequence is probably of relatively shallow water marine origin whereas its upper part and the other sequences suggest much deeper water depositional conditions and the incorporation of ocean floor rocks.

"Cambro - Ordovician"

Sequences of undated, weakly metamorphosed sedimentary and igneous rocks occur within the sheet area and are most probably of Cambrian to Ordovician age (Fig. 7). The Deer Park Complex (DX) of the Westport **inlier**, and the Kill inlier of Clare Island, both include rocks which originated as oceanic crust and even deeper in the Earth's upper mantle. These now consist of metamorphically altered **ophiolitic** rocks, such as serpentinite, **talc**-carbonate schists, metadolerite and metagabbro. Oceanic rocks, similar to some of those in the Deer Park Complex, are thought to underlie all of South Mayo and to have been folded with the Ordovician rocks of South Mayo.

Rocks until recently considered Cambro-Ordovician age, but now interpreted to be of possible Silurian age, are exposed on south Achillbeg and on Bills Rocks (Fig. 7). These rocks include conglomerates, greywackes and black **graphitic** slates, with some quartzites. Folding and two **cleavages** have been recognized.

Two small isolated units in the northeast of South Mayo are of unknown age (Fig. 7). The Farnacht Formation (FN) consists of schistose, moderately metamorphosed volcanic rocks which may be Cambro-Ordovician age. The fairly strongly

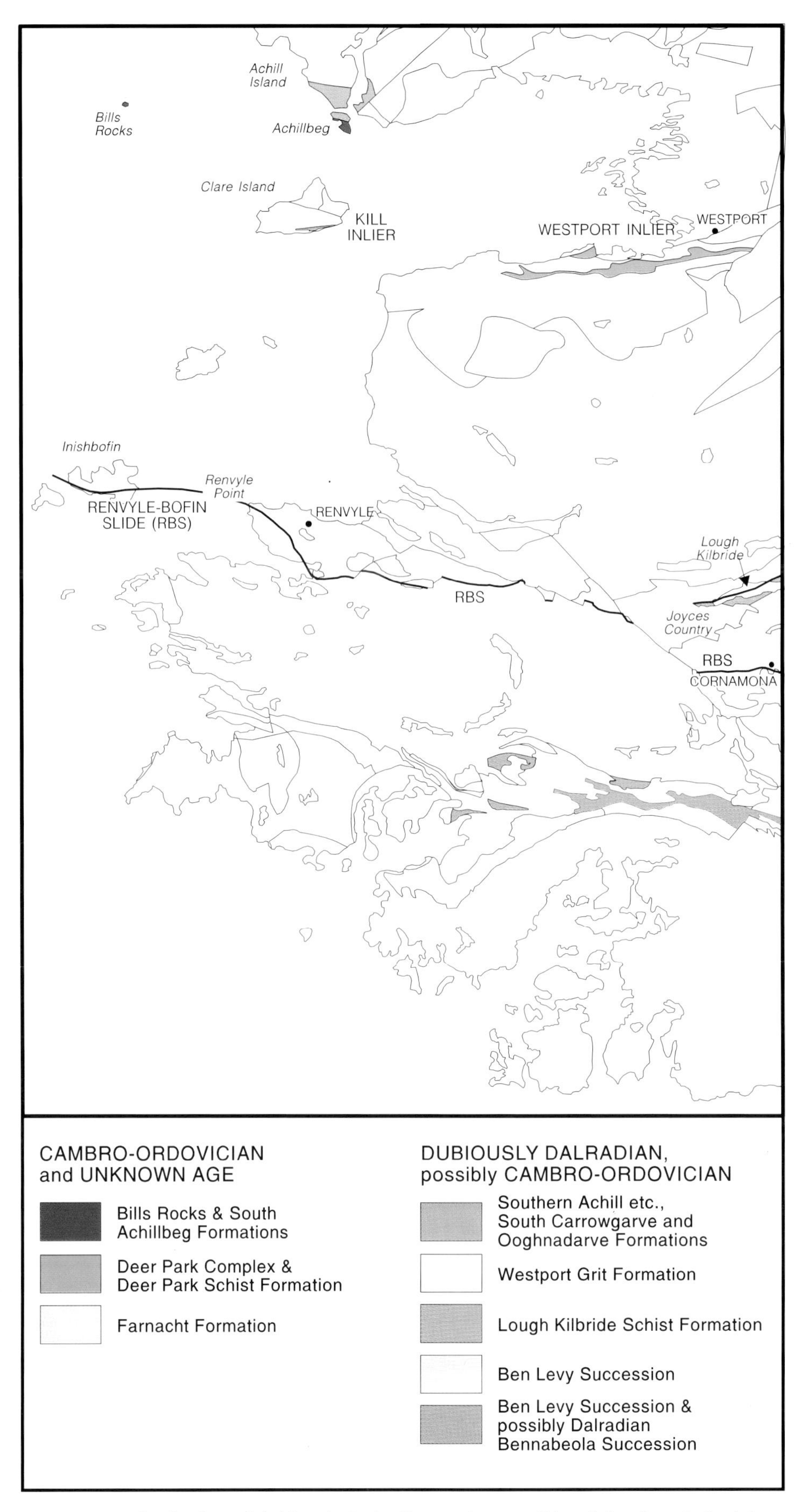

Figure 7. Distribution of dubiously Dalradian rocks, possibly of Cambro-Ordovician age; and "Cambro-Ordovician" rocks.

metamorphosed and repeatedly folded schists of the Deer Park Schist Formation (DP) could be any age between Precambrian (Dalradian or older) to Cambrian or Ordovician.

The Iapetus Ocean: its birth c. 600 million years ago

As noted in the introduction, the later stages of the development of the Dalradian sequence reflect the initiation of the Iapetus Ocean (Figs 4 and 8), the story of which continues throughout and dominates the evolution of rock sequences during the Cambrian, Ordovician and Silurian periods. The ocean opened between what are now North and South America and at its peak width, in the late Cambrian, about 520 million years ago, Iapetus may have been as wide as the present day Atlantic Ocean. And, as if foreshadowing the much later development of the Atlantic, traces of the former existence of Iapetus extend all the way along the eastern seaboard of North America through eastern Greenland, Ireland and Britain into Scandinavia.

Fossil and other evidence indicates that the axis of Iapetus ran through central Ireland, roughly from the present day Shannon Estuary to Clogher Head in Co. Louth, and, rather neatly presaging later historical developments, separated present day England from Scotland. It is difficult now to imagine that during the Ordovician period, present day northwest and southeast Ireland lay along the margins of totally separate continental masses, separated by an ocean several thousand kilometres wide. But that is precisely what one of the "stories written in stone" is telling us and Figure 8 provides a global reconstruction of the positions of oceans and continents during the Ordovician, about 450 million years ago.

At that time, northwestern Ireland, including the Sheet 10 area, lay on the southern margin of the ancestral North American continent, "Laurentia", just south of the equator (Fig. 8). In contrast, southeast Ireland formed part of "Eastern Avalonia", one of two small continental masses located further south in the Iapetus Ocean, at about 30° S (Fig. 8). Eastern Avalonia, and its counterpart "Western Avalonia", were both formed during the rifting and separation of "Baltica" (ancestral Scandinavia) from the ancient supercontinent "Rodinia" (Fig. 4) at about 600 million years ago. Over the next 150 million years subsequent to rifting apart, these two small continental masses gradually drifted northward until they finally collided with the Laurentian margin. The consequences of that collision were profound, as detailed in a later section (p. 27), but, for the moment, we will consider the story told by the rocks which formed before and during that collision: rocks of Ordovician and Silurian age.

Ordovician: 510 - 440 million years ago

Ordovician rocks, 510 - 440 million years old, are exposed in six fault bounded sequences in South Mayo, mainly between Killary Harbour and Clew Bay (Fig. 9). In Connemara, Ordovician sequences include two relatively small areas of meta-volcanic rocks and a regionally extensive suite of metamorphosed igneous intrusive rocks representing the "root zone" of a "magmatic arc" (Main map and Figs 6 and 10).

The rocks record a history of **volcanic arc / island arc** volcanism, deeper level magmatic arc igneous intrusive activity, and oceanic sedimentation, represented by a range of rock types including pillow lavas, tuffs, greywackes, shales and **slates** which accumulated in a small ocean marginal basin known to geologists as the South Mayo Trough. This trough developed during late Cambrian or early Ordovician times, between an offshore chain of volcanic islands (island arc) and the Laurentian continent. The volcanoes are now represented by South Mayo's major groupings of extrusive igneous rocks of Ordovician age near Lough Nafooey, Bohaun and Glensaul (Fig. 9). The development and evolution of the South Mayo Trough is partly recorded in these rocks, but has been deciphered mainly from Ordovician sedimentary rocks.

VOLCANOES

The earliest rocks to accumulate in this small oceanic basin were mainly various types of volcanic rocks erupted from a chain of volcanoes strung along the southern, oceanward side of the basin. They are now exposed in South Mayo, principally in the area around Lough Nafooey, where they are collectively called the Lough Nafooey Group; in the Tourmakeady-Glensaul area; and also in two other areas in Connemara, around Gorumna and just east of Mannin Bay in the "Delaney Dome" (Fig. 9).

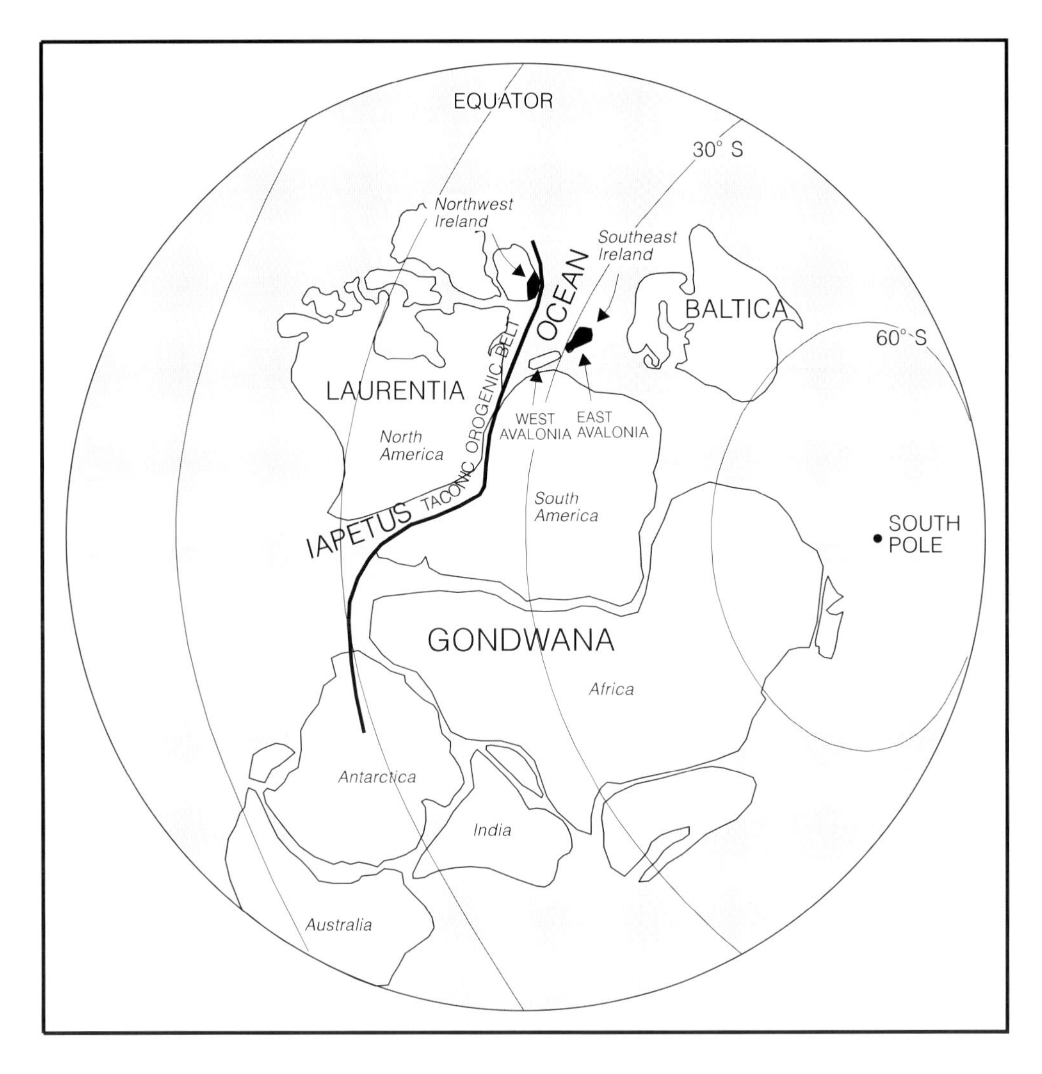

Figure 8. The world 450 million years ago: mid-Ordovician tectonic plate reconstruction showing the Iapetus Ocean (based on Dalziel 1995). The areas of present day northwest and southeast Ireland are shown in black and these have been exaggerated for clarity. Other present day landmasses are indicated by names in italics.

The Lough Nafooey Group is, as shown on the map, subdivided into a number of specific formations. The rocks reflect alternating effusive and explosive volcanic episodes which pass upward through the Ordovician sequence into mudstones, sandstones and fine conglomerates. The earliest, and dominant rock types in the volcanic pile are basalts which were all erupted under the sea. This is indicated by the presence of various diagnostic structures, particularly pillows.

The pillows were generated by molten lava bursting through the consolidated crusts of advancing lava flows, to form a cascade of stacked, interlocking, lava blobs in front of the main flow. Both the pillows, and other forms of basalt, frequently contain small, near spherical or tube-like cavities or infilled cavities called **vesicles** or pipe vesicles in the case of the tubes. These were generated by streams of gas bubbles escaping from the cooling lava. Basalts displaying all these structures are particularly well exposed on the summit of Bencorragh, just south of Lough Nafooey, and also near the intersection of the Finny road and the minor road at Grid Reference (GR) 10035 25935. Other volcanic rocks, such as tuffs and breccias are also present, as well as red and green coloured chert and black mudstones from which fossils of an extinct group of animals, called **graptolites**, have been recovered. It is these graptolites which help us to determine the age of the volcanic rocks.

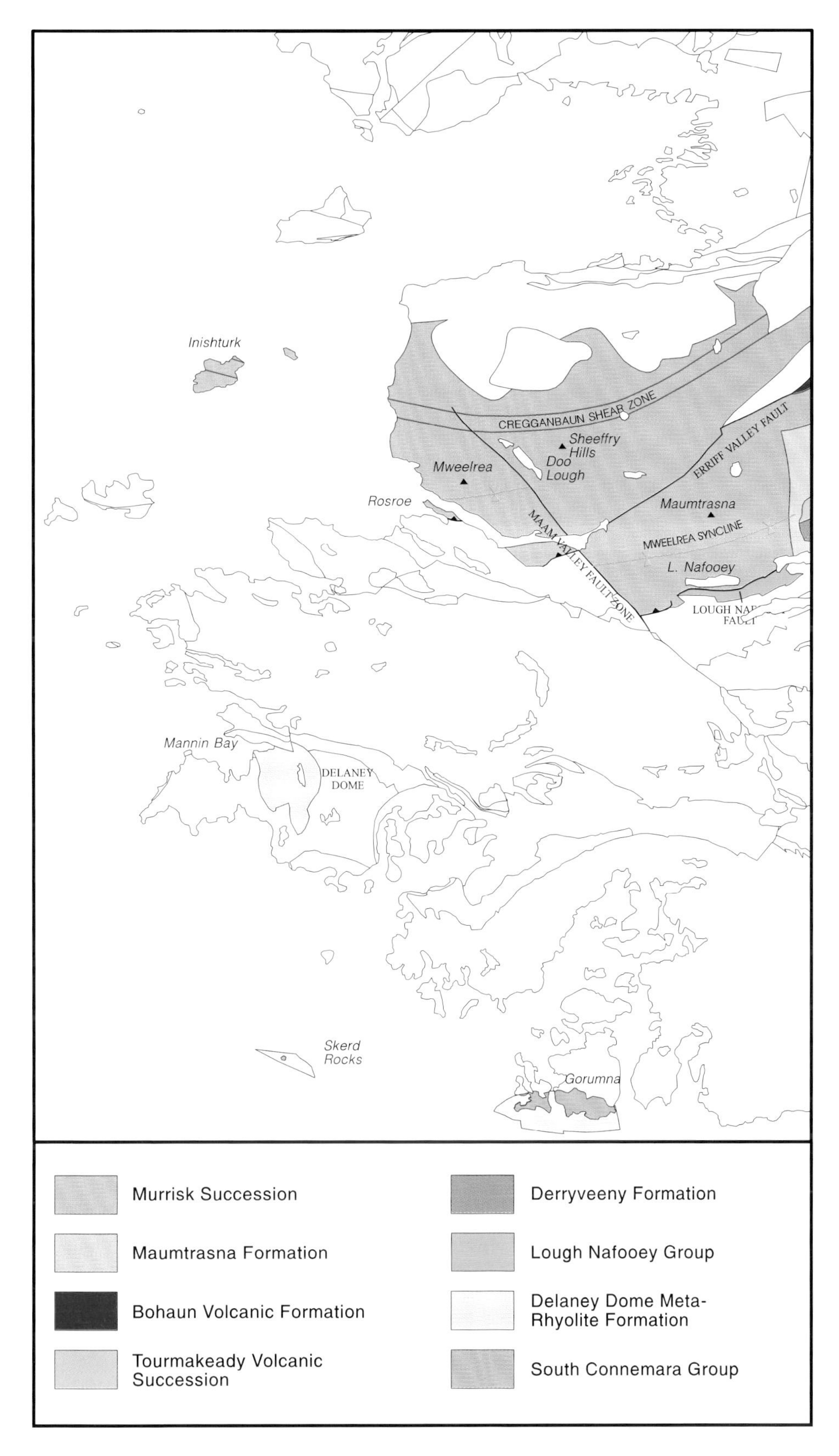

Figure 9. Distribution of Ordovician rocks in the Sheet 10 area.

Plate 5. Early Ordovician (Arenig) basalt pillow lava, Bennane Head, south of Girvan, Scotland. Note the "mushroom" shape of the pillow under the hammer handle (length about 57cm): a markedly convex, upper surface at the tip of the handle and a downward projecting "keel" on the lower, right hand side. The "keel" formed by sinking of the semi-molten lava into a crevice in the underlying surface. This indicates that this pile of pillows accumulated upward from the lower right part of the photo to upper left. Photo by J.H. Morris, 1976.

Plate 6. Intrusive early Ordovician metagabbro (dark green), intruded and broken up by later quartz-diorite orthogneiss. Compare with Mg, Qd and Qg in Figure 6. Hammer handle 57cm long. Errislannan peninsula, southwest of Clifden. Photo by C.B. Long, 1994.

Plate 7. Assorted, well rounded granite boulders in one of the Rosroe Formation boulder conglomerates. Road cutting near Derrynacleigh, approximately 5km west of Leenaun. Photo by J.M. Pulsford, 1981.

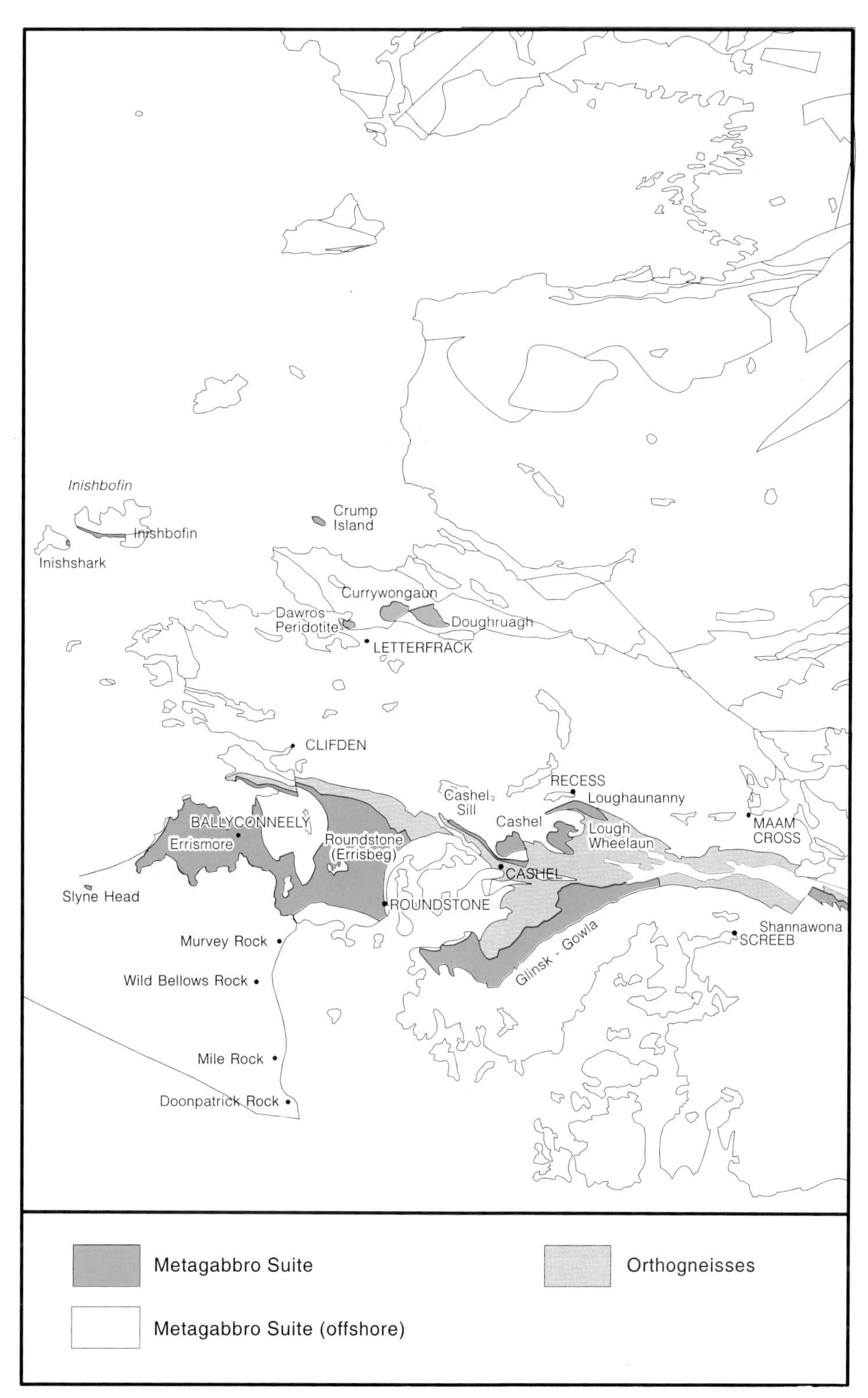

Figure 10. The Early Ordovician magmatic arc-related Metagabbro and Gneiss Complex of Connemara.

The volcanic sequences in Connemara reflect the same general story but differ in detail and particularly their current appearance. In the Gorumna area, the rocks are collectively called the South Connemara Group and consist of a mixed sequence of basaltic lavas, intrusive units and interlayered greywackes, conglomerates and mudstones. They have, however, been strongly metamorphosed within the aureole of a major granite body, the Galway Granite, which occurs along much of the length of the north shore of Galway Bay (see Main map; also Fig. 13).

The situation of the presumed **rhyolitic** volcanic rocks east of Mannin Bay is even more complex as here the highly deformed rocks are exposed in an erosional "window", called the "Delaney Dome", surrounded by Dalradian and arc intrusive rocks (Fig. 6). Although the window rocks are considered to form part of the same volcanic arc sequence, their present configuration indicates that the whole of the Connemara Dalradian block has been thrust, literally pushed en masse, over this part of the arc. It is sobering to reflect on the magnitude of the earth forces which can lift, push and move around huge chunks of the Earth's crust.

In southern Connemara, part of what might be a different arc sequence is represented by large volumes of metamorphosed gabbroic rocks (metagabbros, Mg) and succeeding igneous intrusive rocks ranging from **quartz-diorite** to granite (**orthogneisses**, Qd, Qg), collectively termed the "Connemara Metagabbro and **Gneiss** Complex" (Fig. 10 and Plate 6) and schematically represented by the "magmatic arc" in Figure 6. These rocks, which represent the deeper level, "root zone" of a volcanic arc, were intruded into a continental margin, below active volcanoes, and subsequently fragmented and faulted into discontinuous segments (Fig. 10). The Oughterard Granite may have been intruded either late in the development of this arc or after its formation.

One further, although minor rock type characterises volcanic activity throughout the Ordovician succession, particularly in the Sheeffry (SH), Rosroe (RR) and Mweelrea (MW) Formations. These are explosive volcanic rocks, volcanic ash deposits of various specific types, generally called "tuffs". **Ignimbrites** are a specific type of "tuff" and they represent the fused products of explosively erupted, fluidised admixtures of red-hot volcanic ash and gas. They occur frequently in the Rosroe and Mweelrea Formations and one such horizon is well exposed on the north side of Killary Harbour near Bundorragha (GR 08475 26310), where the road turns sharply away from the harbour towards Delphi and Doo Lough. The ignimbrite occurs in the road cutting on the inside of the bend in the road. On freshly broken surfaces it is slightly shiny in appearance and black, blotched with pink, in colour. The black background represents fused **shards** of volcanic glass, whereas the pink patches represent collapsed fragments of **pumice** which was hot and soft when it came to rest. White flecking in the rocks is formed by crystals of quartz.

BASIN FILL

Kilometre upon kilometre of interlayered sand and mud accumulated on the floor of the South Mayo Trough over a period of about 20 million years during the Ordovician. The sands and muds have long since been converted to solid rocks and compressed into a great, east-west trending syncline, known as the Mweelrea Syncline (Fig. 9).

The oldest Ordovician sedimentary rocks occur at the base of the exposed succession on the northern **limb** of the Mweelrea Syncline. Oldest of all is the Letterbrock Formation (LK) in which conglomerates are interbedded with greywackes and slates. The conglomerates abound with indications of slope-failure and are considered to date from the formative stages of the South Mayo Trough when seismic activity repeatedly triggered slumps, which, broadly speaking, are the submarine equivalents of landslides.

Deposition of the Letterbrock Formation was followed by a prolonged period of deep-water marine sedimentation which is now represented by the Derrymore, Sheeffry and Derrylea Formations (Main map: DM, SH, DL). The three formations are about 5000m thick overall and largely consist of greenish-grey greywackes and slates, formed by turbidity currents, a process described earlier in the Dalradian section (p.15).

The rather monotonous alternation of sandy greywacke beds, ranging anywhere between about 5cm to 3m or more in thickness, and much thinner interlayered slates, imparts a markedly ribbed texture to the landscape. This is well displayed in the Sheeffry Hills range, mainly north and east of Doo Lough. The texture and the organization of the alternations is particularly well displayed along the mountainside road-cut above Sheeffry Bridge (GR 09160 26900) and also in the road-cut by the east side of Doo Lough.

Towards the top of the turbidite sequence, the deep oceanic conditions were gradually replaced by increasingly shallow water conditions as the basin filled. The start of this infilling process is evident in both the Derrylea and Rosroe Formations (DL, RR). The latter consists of a variety of rocks, including thick horizons of ignimbrites and, most notably, of conglomerates which contain very well rounded off cobbles and boulders up to 1m in diameter (Plate 7).

The appearance of these conglomerates for long led geologists to infer that they represented ancient beach deposits. Nowadays, the formation is considered to reflect deep sea deposition, at the mouth of a submarine canyon which cut across the steep, southern margin of the South Mayo Trough.

The Rosroe Peninsula is situated at the southwestern extremity of the South Mayo Ordovician outcrop (Fig. 9). Towards the eastern edge of the outcrop, in the Partry Mountains, an extensive area is underlain by conglomerates and sandstones not unlike some of those in the Rosroe Formation, but of different origin and controversial age. These rocks, referred to as the Maumtrasna Formation (MT), are thought to have originated in the **braided** channels of **alluvial** fans in the borderlands of the South Mayo Trough. Whether the alluvial fans co-existed in time with the deep-sea submarine environments represented by the Rosroe and Derrylea Formations, or whether they co-existed with the shallow marine conditions which post-dated the deep-sea stage in the history of the South Mayo Trough, remains unresolved.

The deposition of the Derrylea Formation and the Rosroe Formation marked the beginning of the end of deep-sea sedimentation within the South Mayo Trough. Deposition of the two formations was followed by a period of reduced sediment supply to the trough and the widespread deposition of submarine muds, now represented by the slates of the Glenummera Formation (GU). In the uppermost levels of this Formation, thin beds of **flaggy** sandstones are interbedded with the slates and mark the onset of shallow water across the site of the former trough. However, as basin subsidence kept pace with infilling, the upper part of the succession, exposed in the mountain range stretching from Mweelrea in the west to Maumtrasna in the east, consists of over 4,000m of shallow water reddish coloured sandstones and conglomerates of the Mweelrea Formation (MW).

Silurian: 440 - 410 million years ago

Like the Ordovician rocks of the region, the Silurian rocks of South Mayo form part of the rocky foundations of the mountainous country west of Lough Mask, where two separate tracts of Silurian rocks are exposed. One tract occurs along the north flank of the Mweelrea Syncline, the other to its south (Fig. 11).

The northern tract of Silurian rocks includes those of Clare Island and others which extend obliquely inland from the southwestern corner of Clew Bay to the vicinity of the Bay's landward extremity (Fig. 11). This tract also includes the bare conical quartzite peak of Croagh Patrick. The southern tract lies largely in north County Galway and extends inland from the coast south of Killary Bay Little, across the mountains of Joyces Country to the vicinity of Lough Nafooey.

Whereas the Ordovician rocks were deposited offshore from the south coast of Laurentia, the Silurian rocks of the South Mayo region were deposited both on the continental margin, as well as beneath adjoining shallow to deeper water areas (Fig. 8). However, much had changed there during the interval between the end of Ordovician sedimentation about 460 million years ago and the commencement of Silurian deposition about 430 million years ago.

Late Ordovician earth movements buckled South Mayo's Ordovician rocks into broad, open folds and shuffled the Dalradian rocks of Connemara and the South Mayo Ordovician rocks sideways several hundred kilometres against each other. The two sets of rocks are believed to have migrated from totally different parts of the Laurentian margin area and to have been assembled into approximately their current configuration before the formation of South Mayo's Silurian rocks.

Gone were the volcanic islands of Ordovician times and with them most of southern Laurentia's active volcanoes. In their place now stood jagged, coastal mountain ranges formed of newly uplifted Dalradian and Ordovician rocks. As early Silurian erosion took its toll of the ranges, pockets of sedimentation developed and some of them became fully fledged depositional basins.

Three such basins are represented in the Silurian rocks of the South Mayo region (Fig. 11): the **littoral** to non-marine sediments of the Louisburgh - Clare Island Succession; the shallow marine

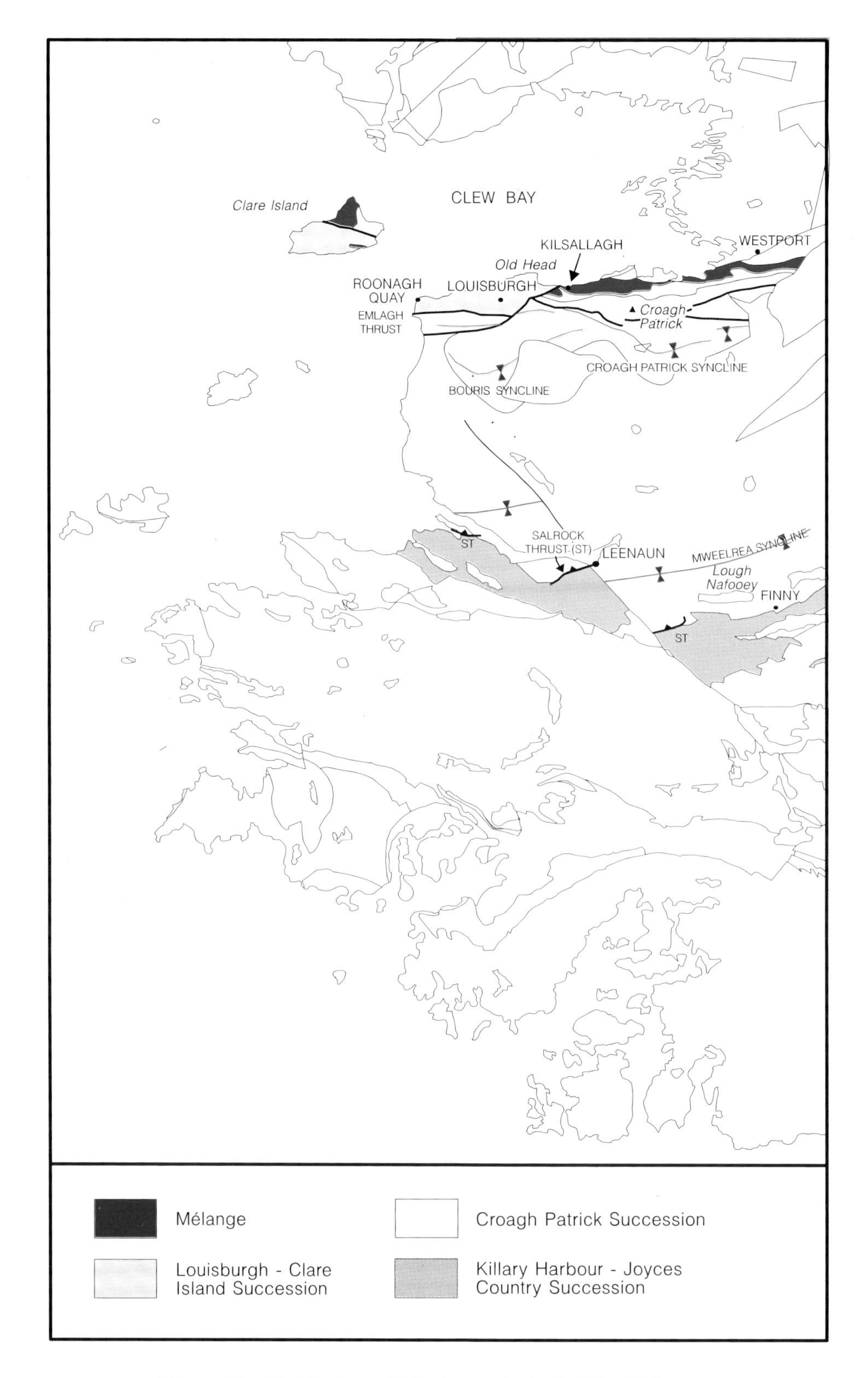

Figure 11. Distribution of Silurian rocks in the Sheet 10 area.

deposits of the Croagh Patrick Succession and the **fluviatile** to deep marine deposits of the Killary Harbour - Joyces Country Succession. A fourth sequence, only very recently determined to be Silurian age, is exposed on Clare Island and along the south shore of Clew Bay. This sequence consists of a somewhat unusual assemblage, a mélange, of black shales containing large, sometimes very large, sandstone and chert blocks which originated by soft sediment slumping on an unstable slope.

LOUISBURGH-CLARE ISLAND SUCCESSION

The 1000m thick succession occurs along the southwestern shores of Clew Bay from Old Head to Roonagh Quay and on Clare Island (Fig. 11). On the mainland, the succession is separated from older rocks of the Silurian Croagh Patrick Succession by a major, low-angle reverse fault, known as the Emlagh Thrust (Fig. 11).

Red, grey or green siltstones and mudstones, along with grey, greenish-grey, or buff-coloured sandstones are the principal kinds of rock in the succession, although conglomerates and pebbly sandstones occur at the base of the succession and again at the top. Ripple marks and desiccation cracks are common and overall, the succession is considered to have originated as mud-flats on the floor of an ephemeral, **playa** lake in a land-locked, semi-arid, non-marine environment.

CROAGH PATRICK SUCCESSION

The Croagh Patrick Silurian Succession, which occupies the ground south of Clew Bay (Fig. 11) contains none of the brightly coloured siltstones so characteristic of the Louisburgh - Clare Island Succession. Instead, it consists largely of varieties of grey sandstone, or, more correctly speaking, of lightly metamorphosed grey sandstones (psammites and quartzites), marbles and siltstones containing a sparse but wide variety of fossils such as **corals**, **trilobites** and **brachiopods**. The entire assemblage reflects deposition in a shallow marine environment.

Part of the succession forms one of South Mayo's best known, and rockiest landmarks, Croagh Patrick. However, most of the rocks which make up the succession are obscured from view by the great expanse of low level blanket bog south of Louisburgh, or by glacial moraines in the country west southwest of Westport.

KILLARY HARBOUR - JOYCES COUNTRY SUCCESSION

Throughout its extent, the Killary Harbour - Joyces Country Succession rests with angular discordance either on the possibly Dalradian rocks of the Connemara **Massif**, or on rocks of the South Mayo Ordovician outcrop.

The succession consists of a 2.5 km thick pile of folded and faulted sedimentary rocks which record evolution from initial non-marine conditions, through continental shelf sedimentation to oceanic depths and, finally, back to shallow water, shoreline environments.

The succession typically begins with a layer of **breccia** composed of angular fragments of older rocks. This is overlain by river-lain coarse grained, purplish-red sandstones, derived from the coastal mountain ranges which lay along the northern margin of the Iapetus Ocean. This margin was gradually submerged beneath Iapetus and this is reflected in the rock record by a transition upward into grey, marine, fossil-bearing sandstones and 1,500m of monotonous, graded, grey turbidites.

The turbidite succession marks the maximum northward encroachment of the Iapetus Ocean onto the Laurentian continental margin during the Silurian. The remainder of the Killary Harbour - Joyces Country Silurian Succession was formed under progressively shallower, marine conditions, as now represented by cross-bedded sandstones and red and green mudstones at the top. These represent sedimentation in extensive shoreline mudflats, which were sparsely populated by one of the world's longest surviving genera, the brachiopod *Lingula*, which is known from rocks more than 500 million years old.

Destruction of the Iapetus Ocean, 440 - 390 million years ago

(See also Harley *et al.* 1990, "Death of an Ocean".)

By the end of the Silurian period, the 200 million year reign of the Iapetus Ocean as a dominant feature of the ancient environment was nearly over. Figure 12 presents a reconstruction of the configuration of the world's oceans and continents after completion of the closure of the Iapetus Ocean. Prior to that, opposing continental margins had been slowly creeping towards each other throughout most of the Ordovician and Silurian periods, at maybe

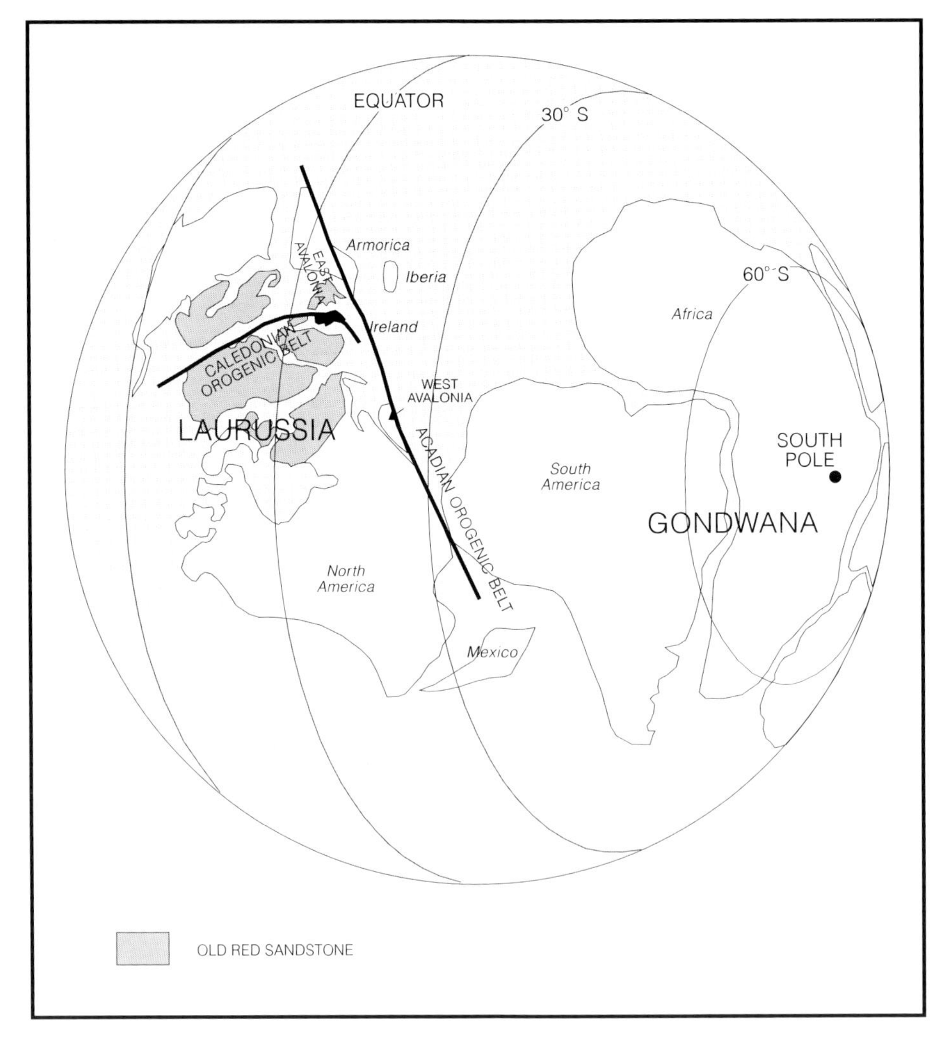

Figure 12. The world 370 million years ago: a Devonian tectonic plate reconstruction, based on Dalziel (1995). The area of Ireland is shown in black.

no more than a centimetre or so per year, the rate at which the modern day Atlantic is currently spreading. The convergence reflects consumption of oceanic crust beneath the opposing margins of Avalonia and Laurentia. When the opposing continental masses finally collided is a still a matter of debate, some arguing that it did not occur until the end of the Silurian or early Devonian. **Subduction** of oceanic crust had, however, largely, but not totally, ceased by the end of the Ordovician.

Final closure took place during late Silurian times and resulted in strong deformation of the rocks caught up between the two colliding continents: cascades of folds were generated, much like rucks in a carpet, with successive generations of folds refolding earlier folds, while fractures and faults cut and displaced rock sequences in all directions. In South Mayo and Connemara, this is reflected by a progressive succession of distinct deformation effects, including tightening of the Mweelrea Syncline; intense folding of the turbiditic formation on the northern limb of the syncline, particularly in the Cregganbaun Shear Zone (Fig. 9); displacement on the Maam Valley fault zone; and **overthrust**ing of the Ordovician rocks south of Killary Harbour, along the Salrock Thrust (Fig. 11), onto the younger Silurian strata.

Just as the collision of India with Eurasia, in geologically much more recent times, has formed the loftiest mountain range at present, the Himalayas, so the final stages of the end-Silurian/ early Devonian continent-continent convergence raised a mighty mountain chain, possibly of equally Himalayan proportions. From the Appalachians in

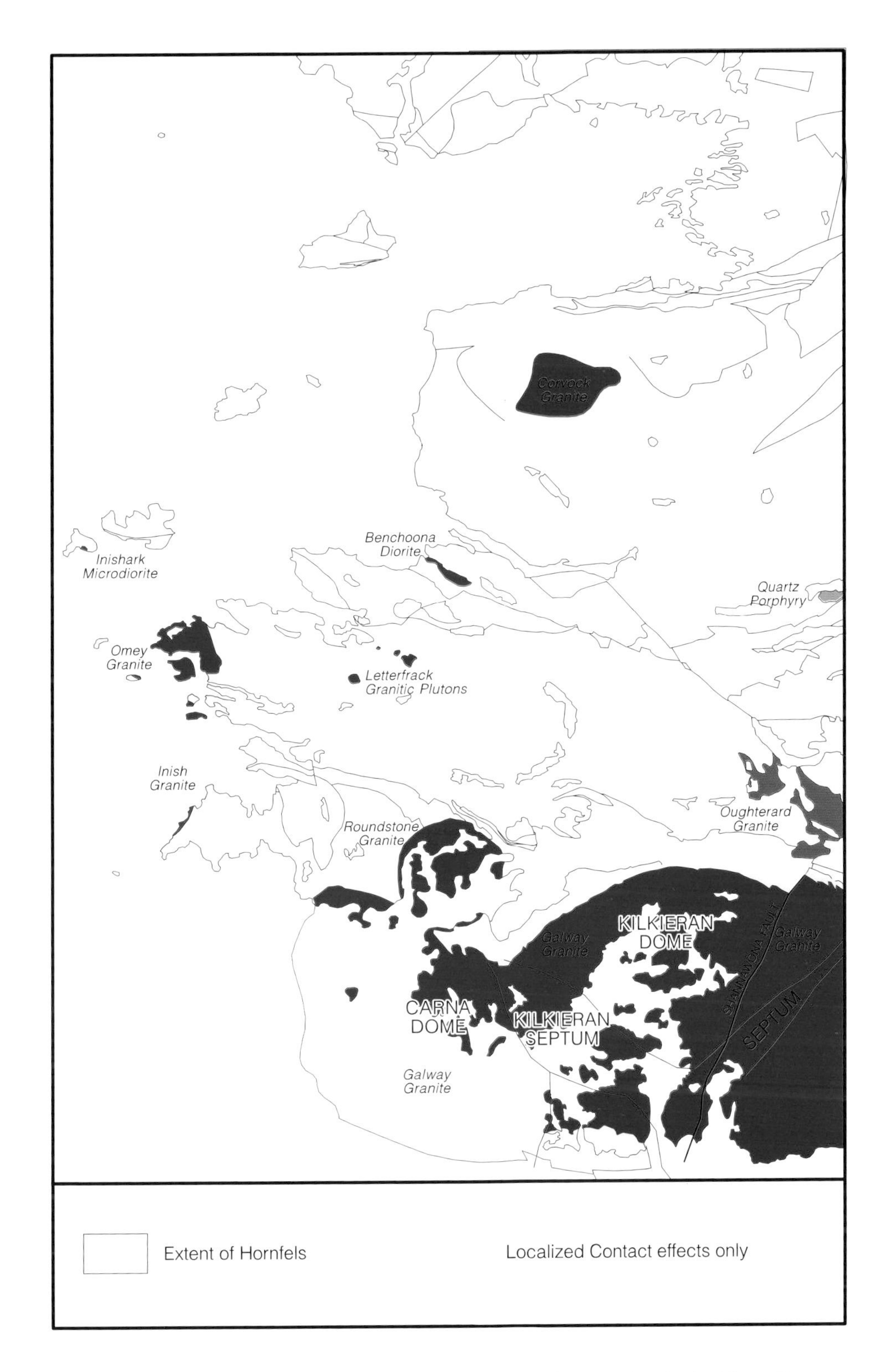

Figure 13. Distribution of end Silurian to early Devonian granites in the Sheet 10 area.

North America to the Arctic Circle in Norway, remnants of this ancient Appalachian-Caledonian mountain belt are to be found today in the regional Precambrian, Cambrian, Ordovician and Silurian rocks, including those of the South Mayo region. These are the cataclysmic scale forces, resulting from continent-continent collision, which were able to convert the deep-ocean sedimentary rocks exposed today, for example in the Sheeffry Hills, into mountains and push and shove a huge block of Dalradian rocks over part of the Ordovician volcanic arc within the Delaney Dome.

As rocks were pushed up to form mountains, some were also pushed down to great depths where the heat and pressure caused them to melt, producing large volumes of molten granite magma. This relatively light, buoyant magma rose upwards, intruded into the upper crust, where it cooled slowly to form coarse grained granite. Several granites of early Devonian (late Caledonian) age were intruded in Connemara and constitute the large Galway Granite **batholith** and its associated satellite **plutons** (Figs 6 and 13). The Corvock Granite of South Mayo is another body of this general age. Granite intrusion was generally widespread at this time and might be related to the very last subduction movements following closure of Iapetus. Intrusions are geographically associated with major and fundamental fault lines or shear zones.

The hills and mountains of the Appalachian-Caledonian mountain belt have long since been eroded away, mainly during the Devonian and early Carboniferous periods, and the only vestiges we see now are the rocks which once lay deep down in the roots of those mountains. By the mid-Carboniferous, the mountains had been largely levelled by erosion, in the process providing huge volumes of mud, sand, gravel and pebbles to build a new generation of rocks - a massive and entirely natural recycling process!

The global reconstruction presented in Figure 12 provides an outline of the disposition of the world's oceans and continents during the mid-Devonian, about 370 million years ago. Comparison with Figure 8 clearly illustrates the dramatic changes in configuration which had occurred throughout the intervening 80 million years. The Iapetus Ocean, and related seas, had been completely destroyed by the collision of the Avalonian micro-continents and Baltica with Laurentia to form a new continent "Laurussia". One misleading aspect of these reconstructions is that Figures 8 and 12 create the impression that during that time interval Laurentia remained attached to Gondwana. In fact, Laurentia had completely detached from Gondwana by about 590 million years ago and reconverged with it twice at about 450 and 370 million years ago. From then on, the world was on the way to consisting of a single global ocean and a single supercontinent, "Pangaea", which would, however, still take another 100 million years to construct.

At the more local level, compare the position of southeast Ireland in the two diagrams: it had migrated from a latitude of 30°S during the mid-Ordovician to a near equatorial position by the mid-Devonian. Northwest Ireland, including "Connemara", had also migrated closer to the equator, but did not travel anywhere near as far as southeast Ireland. Certainly by the beginning of the Devonian, the land area that was to become modern day Ireland had been amalgamated into a single land area, on the southern, equatorial margin of Laurussia (Fig. 12). That position remained more or less static throughout the succeeding Devonian and Carboniferous periods, to which we will now turn and continue our story.

Devonian (Old Red Sandstone): 410 - 360 million years ago

Continental collision wrought a dramatic change to the environment of the area that was to become Ireland. Ocean was transformed into land, as bits of what was to become Ireland had migrated from a mid-latitude position to join the rest of Ireland almost on the equator (Fig. 12). Given such a location, it should come as no surprise as to what the dominant environment was throughout the Devonian: a desert, or more strictly speaking, an arid landscape of mountains, hills and plains subject to periodic, torrential rainfall which generated transitory rivers and lakes. The ephemeral lakes and rivers of South Australia and the bare rock uplands of Western Australia provide good modern day analogies to that long since vanished Irish environment.

The rocks which record this phase in the environmental evolution of Ireland are very poorly represented in this particular map area and consequently a fuller account of that story is best left to those areas where they are well represented, in Counties Cork and Kerry.

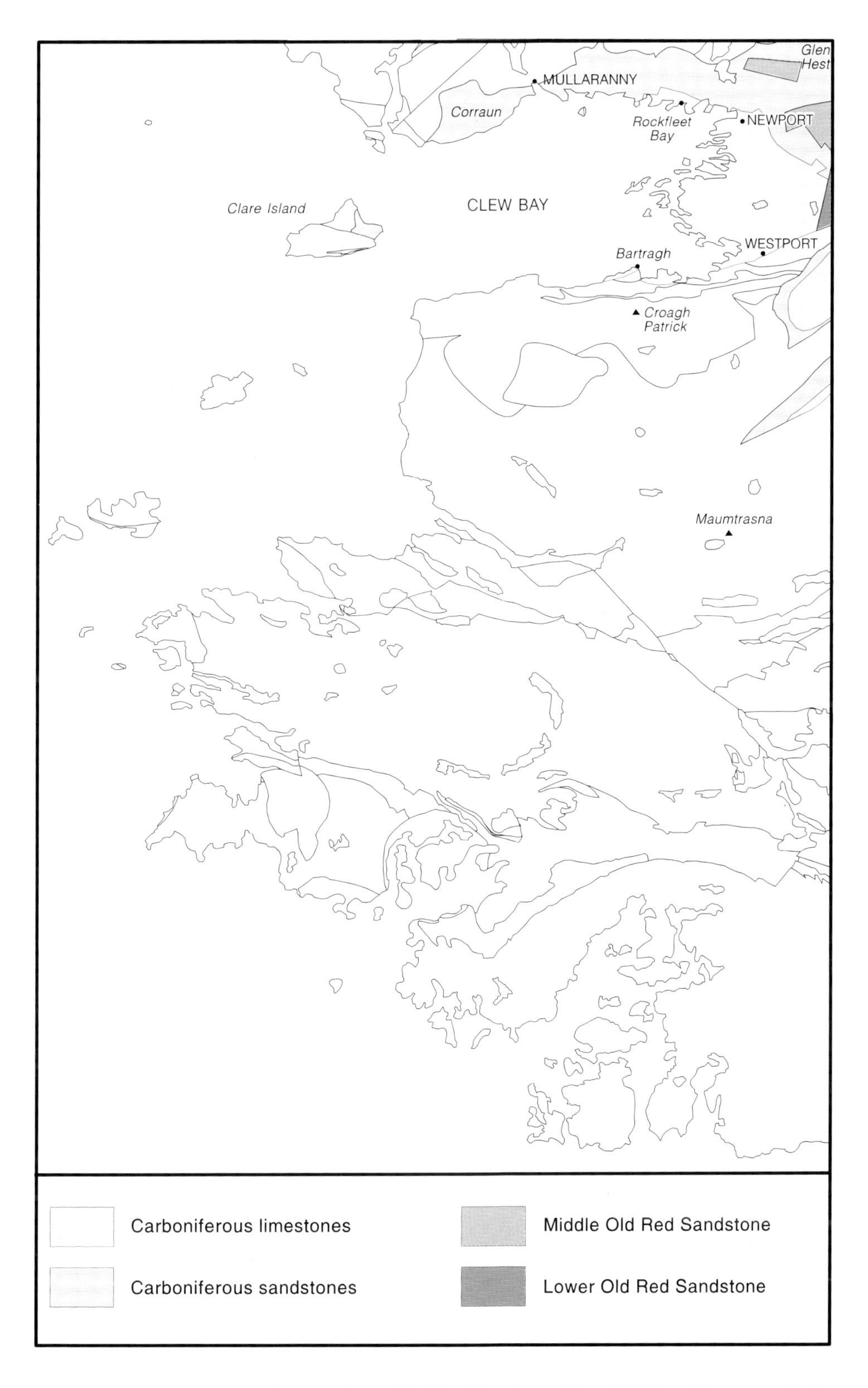

Figure 14. Distribution of Devonian ("Old Red Sandstone") and Carboniferous rocks in the Sheet 10 area.

These **Old Red Sandstone** deposits (Fig. 14) reflect progressive erosion of the Caledonian mountains created by continental collision. The transitory rivers carried vast volumes of material onto plains fringing the mountains, depositing the coarser materials on alluvial fans close to the mountains, and sand and mud in great sheets on the plains or in the ephemeral lakes. Those deposits are now represented by generally reddish-brown coloured conglomerates, sandstones and siltstones within both the Lower and Middle Devonian sequences on the map. The Lower Devonian rocks are in faulted contact with, or rest unconformably on older rocks, whereas the Middle Devonian sequence is more widely distributed, and is faulted against Lower Devonian rocks. The latter rocks were deformed by a single phase of folding prior to the deposition of the Middle Devonian rocks and both were then effected by a further episode of folding prior to deposition of the overlying Carboniferous rocks.

Carboniferous: 355 - 290 million years ago

Carboniferous age rocks underlie more than half of the present day land area of Ireland, principally in the central plains, stretching from east Galway to Dublin, north to Sligo and Donegal, and away south to Counties Tipperary and Limerick. The story of these rocks is consequently of perhaps more importance to the building of Ireland than that of any other rocks, but, as there are, again, very few exposures of such rocks in the Sheet 10 area, that story is best left to be told elsewhere. Suffice to say here that the Carboniferous period reflects the gradual northward inundation of the edge of the Devonian continent by shallow tropical seas, a process which took about 10 million years to cover most of Ireland.

Rivers continued to flow southward, dumping sand and other coarse material close to shore, while finer mud and silt were deposited further offshore. This pattern is marked, in a very general sense, by upward and lateral transitions between fluviatile, deltaic and near-shore marine sandstones, and thick mudstones in offshore basins. The intervening, shallow water platform was the site of formation and accumulation of the many different types of limestones and reef-like rocks which so typify the Carboniferous in Ireland. The transition from desert to tropical sea is also marked in a much more obvious way: the desert sandstones contain scant, if any, fossil remains of any type, in stark contrast to the succeeding limestones and mudstones which contain very rich assemblages of marine fossils, the skeletal remains of what were once seas teeming with life.

Carboniferous rocks within the Sheet 10 area are mainly restricted to the area around Westport, and in a few smaller areas to the south (main map and Fig. 14).

SANDSTONES

Carboniferous age sandstones and conglomerates belonging to the later stages of the Old Red Sandstone semi-arid depositional environment rest unconformably on older rocks, and were deposited by southward flowing rivers near the coast. They are represented by the Maam Formation (MM) of Clare Island and the northern side of Clew Bay. Changing climatic conditions are reflected by the mostly grey sandstones and siltstones of the overlying Capnagower and Moy Sandstone Formations (CP, MO), which were deposited by rivers on the coastal plain.

LIMESTONES

Dark-grey, fine-grained limestones interbedded with shales typify the Rockfleet Bay and Castlebar River Limestone Formations (RF, CR) which mark a change to marine conditions and deposition on the landward side of a carbonate shelf.

The overlying limestones are typified by inter-bedded limestone and subordinate dark-grey shale. Upwards, the beds become cleaner and shale decreases, reflecting deposition further away from the land. An episode of nearby delta building, perhaps to the northwest, is represented by the **oolites** and calcareous sandstones of the Westport Oolite.

Igneous Intrusions (Carboniferous and later)

Carboniferous, and particularly Tertiary age, intrusions in the Sheet 10 area all relate to the eventual opening of the Atlantic Ocean about 65 million years ago. The intrusions consist of various types and forms of dolerite, the distribution of which is summarised in Figure 15. Carboniferous age dolerite dykes in Connemara and South Mayo reflect an early, failed rifting event that predates subsequent rifting, from Triassic times onward, which finally led to opening of the north Atlantic Ocean in late Cretaceous and early Tertiary times.

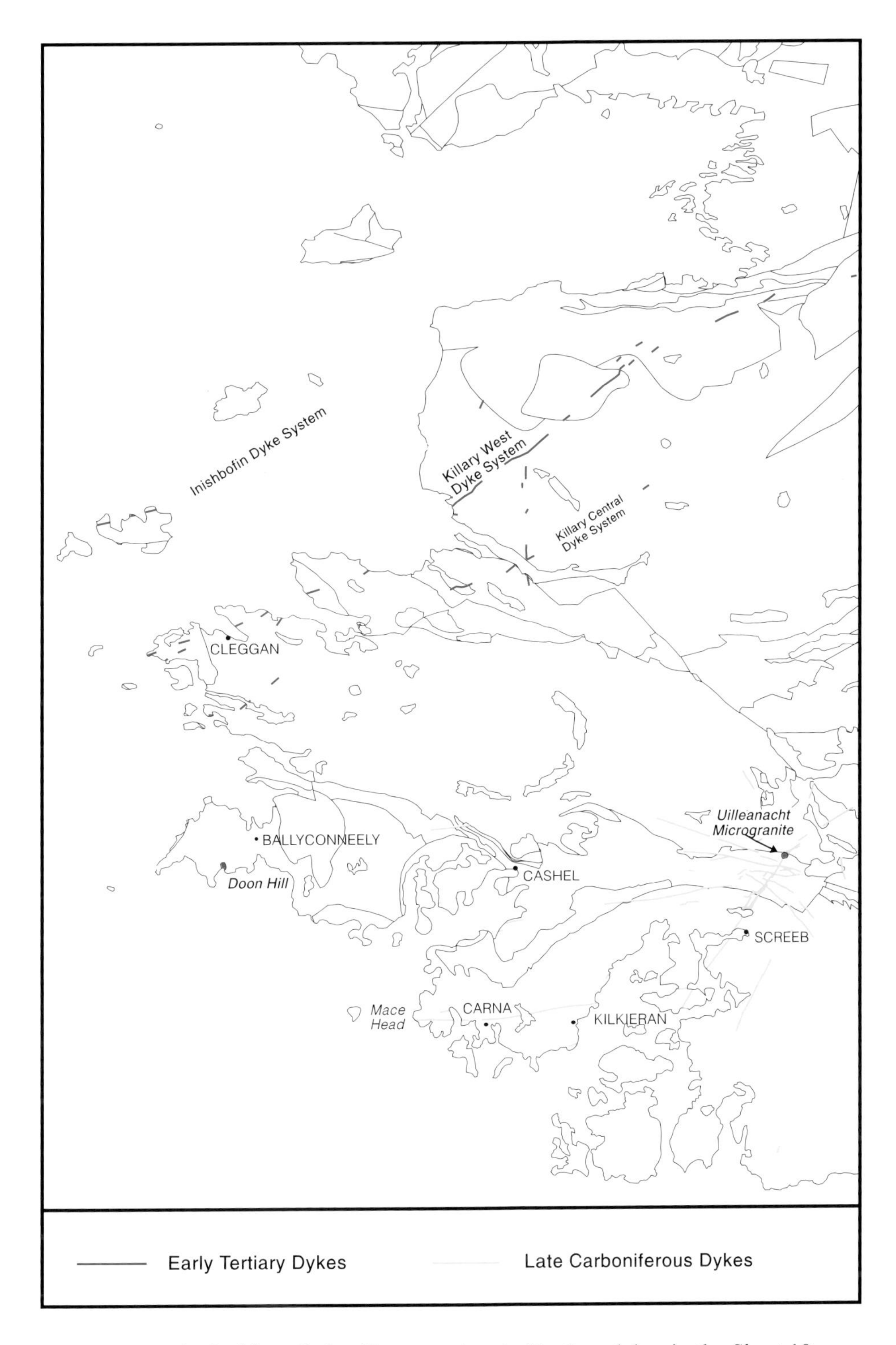

Figure 15. Principal late Carboniferous and early Tertiary dykes in the Sheet 10 area (based on Mohr 1993 and Mohr 1982 respectively).

Tertiary intrusions include several north-east trending dolerite dyke systems which have been recognized in Connemara and South Mayo (Fig. 15). This age group also includes a small **olivine** dolerite "plug" which forms the prominent Doon Hill near Ballyconneely in southwestern Connemara, and a similar plug, the Siofra (Sheeffry) Gabbro in South Mayo.

Quaternary

(See also Nield *et al.* 1989, "The age of Ice".)

Most of the bedrock of Ireland is covered by more recent, generally unlithified, sediments of the Quaternary Period. This Period was characterised in Ireland by alternating cold stages, during which much of the country was covered by great ice sheets, and warmer **interglacial** stages, during which climatic conditions were much as they are now. The Quaternary Period extends from 1.6 million years ago to the present day, and the last cold stage came to a close 10,000 years ago.

Although glacial sediments, which are sometimes very patchy, dominate the Quaternary deposits of the area of this map, they are, for the most part, masked by widespread postglacial peat. The glacial sediments were deposited during the last cold stage of the Quaternary by the great ice sheet which then covered almost the whole of Ireland. Earlier glacial and interglacial deposits were either removed by this ice sheet or buried by its deposits.

The Irish ice sheet was composed of a number of ice domes which expanded from local centres finally to form an extensive ice sheet with a complex flow pattern (Warren, 1992). The area of this map was dominated by an ice dome which was originally centred close to the coast or in the offshore shelf area. It migrated inland until it was centred on a line extending from Clew Bay to Slieve Aughty in County Galway. The thickness of this ice-dome can be gauged from the fact that it overtopped both the Beanna Beola mountains and Maumtrasna mountain. Another indication of its extent is the fact that glacial **erratics** from Connemara have been traced as far east as Slieve Bloom.

During glaciation, the ice sheet eroded the underlying bedrock in places by picking up pieces of loose debris which adhered to the base of the ice sheet and by grinding down the bedrock over which it passed. In this way the rock surface became polished, scratched and moulded by the ice, thereby generating the **striae**, roches moutonnées and generally ice-sculpted landscape that is particularly characteristic of southern Connemara. Indeed, with the exception of Murrisk and the low ground at the head of Clew Bay, between Westport and Newport, such features are common throughout the area of the map, and provide striking evidence of the flow direction of the ice sheet (Fig. 16).

The debris which became entrained in the ice was eventually deposited either directly by the ice as **till** (boulder clay) or by glacial meltwater as clay, silt, sand or gravel. Although Connemara is characteristically an area of glacial erosion, there are extensive tracts of glacial deposits which often take the form of drumlins. These are particularly common in the coastal area between Ballyconneely and Renvyle and on the low ground close to the western part of the N 59. Thick glacial deposits also occur in the Murrisk area, north of Killary Harbour. However, it is on the low ground around Clew Bay that the most spectacular glacial deposits occur (Fig. 16). These are generally tills which take the form of drumlins, some of which form the small islands which dot the head of the bay. The drumlins onshore are noteworthy from two points of view: they are particulary long and narrow (often in excess of 3km long and generally less than 500m wide), and their trend is often curved, running from west to east and then from south to north. These were formed by ice flowing from west to east out of Clew Bay and then deflected to the north by ice flowing from a second ice dome in the north midlands.

Well sorted sands and gravels are not common in this area. There are some glacial meltwater gravels in the form of an **esker** at Tullywee, between Letterfrack and Kylemore. There are also delta sands and gravels at Tullywee and just south of Leenaun and there are considerable spreads of sands and gravels southwest of Erriff Bridge in the Erriff valley. On the southeastern slopes of the valley, there are thick **kame** terrace deposits, and **kame and kettle** gravels occur on the valley floor northeast of the road between Erriff Bridge and Srahlea Bridge, and as far northeast as Derrinkee bridge.

Corrie (or cirque) basins are common in some of the mountains of the area. These were the last areas to be occupied by glacial ice during the final part of the last cold stage, and many contain **moraine** ridges which mark the margins of these glaciers.

A very large part of the area of this map is covered by blanket bog, particularly in Connemara and Murrisk (Fig. 16). This includes both low level "Atlantic" type and high level "montane" type, but the Atlantic type is most common and covers most of the low ground of Connemara. Blanket peat is generally thought to have begun to develop at a period of minor climatic deterioration about 4,000 years ago (see Hammond, 1981).

Other postglacial or Holocene deposits include river **alluvium**, beaches and blown sand, all of which continue to form today. Many of the beaches of

Connemara are composed of carbonate sand (Bosence 1979), some being formed from broken fragments of the coralline red alga *Lithothamnium*, others of finely comminuted debris of bivalves and foraminifera. The beaches around Mannin Bay, southwest of Clifden, are particularly noteworthy, and the coral strand at Ardmore, between Carna and Kilkieran, is well known.

Likely Late Pleistocene faulting has been investigated by Mohr (1986).

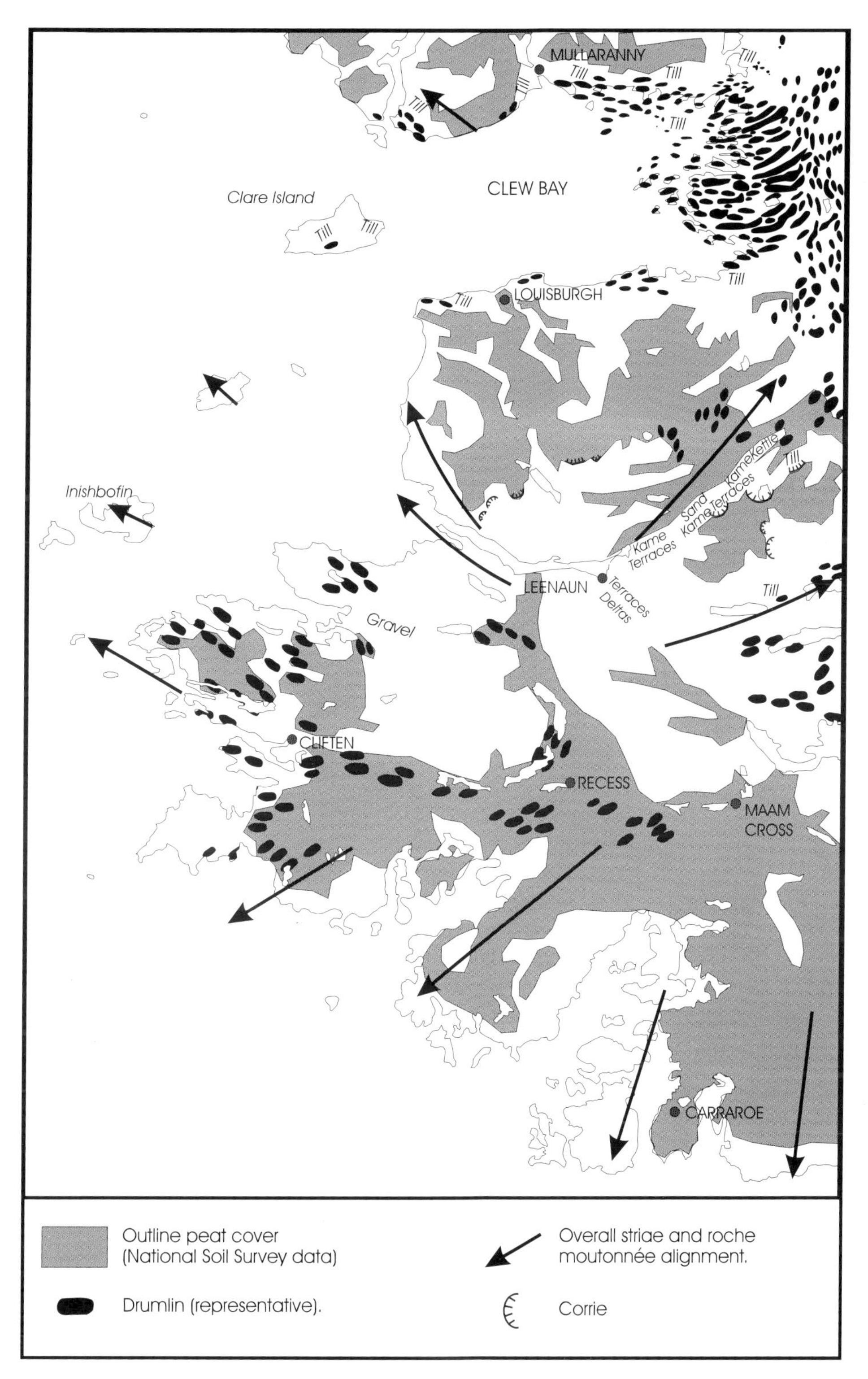

Figure 16. Map showing some of the features of the Quaternary geology of the Sheet 10 area. Peat distribution modified from Hammond (1981).

THE EARTH RESOURCES OF CONNEMARA

MINERAL DEPOSITS

The complex geological history of the map area is reflected by the presence of a rich and diverse range of mineral deposits. Only the principal known deposits are discussed here, and these in only a summary manner. Some technical terms are unavoidable in this section, although some are explained in the text, including the composition of metallic minerals, where first mentioned. The chemical symbols for the metals, such as Au for gold, are used frequently for convenience, particularly in tables, and indicated in parentheses after the metal name in the title of each type of deposit.

The descriptions of the mineral deposits occurring on this map sheet rely on the published work of the Geological Survey of Ireland (GSI), the published work of others (referenced as appropriate) and upon unpublished GSI database records. No new research or site investigation was carried out in compiling this summary. The deposits described here are only those which have been either active in the past; those which represent significant potential for development or those which have been well documented. The numbers in brackets after the deposit name in the text below, and after the deposit symbol on the map, are those assigned to that locality in the GSI Minerals Locality Database. There are very many further less well documented occurrences which have been omitted from the map sheet. Information on these may be obtained from the GSI.

Metallic mineral deposits

Noteworthy metallic mineral deposits are described generally by reference to their principal metal(s) content, as follows: gold; lead-zinc-copper; copper-molybdenum; iron-copper-moybdenum-tungsten; "skarn" deposits; copper-nickel, platinum-palladium and chrome; iron; and baryte. Exploration, and development in some instances, has evolved from a focus upon "base metal" (lead, zinc, copper) deposits up to the early part of this century, to a more recent focus upon such metals as copper, molybdenum, tungsten, platinum, chrome and, in particular, gold since the 1970s.

GOLD DEPOSITS

Gold was first discovered in the Sheet 10 area by Tara Prospecting Limited in 1984, as part of a systematic exploration programme in northwest and western Ireland. Since that initial discovery, at Lecanvey on the south side of Clew Bay, several other significant discoveries have been made, particularly in the Cregganbaun area south of Louisburgh. Gold deposits have now been defined in and around Lecanvey (localities 5513 and 4414), near Croagh Patrick; west of Doo Lough in South Mayo; on Inishturk Island; and at Bohaun (2321) south of Lough Kilbride;

The Lecanvey discovery of 1984 provided the focus for subsequent exploration work by the Tara Prospecting Limited - Burmin Exploration and Development Plc joint venture during the period 1987 to 1990. The prospecting licence within which the deposits are located was not renewed by the Minister for Energy in 1990.

An excellent account of the geology, prospecting and controls on the mineralization is presented by Aherne *et al.* (1992). Gold occurs in quartz veins within the Cregganbaun Formation. Two main zones of mineralization have been identified - the Brocaigh Zone and the Garrymore Zone. The average width of gold bearing veins in both zones is 1m.

Gold occurs with the metallic sulphide minerals pyrite, or "fools gold" (iron sulphide), pyrrhotite (another form of iron sulphide), and chalcopyrite (copper-iron sulphide); niccolite, a nickel arsenic mineral; and iron oxides (haematite) and iron hydroxides (goethite). These minerals themselves form a relatively minor proportion of the rock mass. The margins of the veins are often the loci for the concentration of visible gold which is relatively coarse grained (10 microns to 3mm), with a gold to silver ratio of 9 to 1 (Aherne *et al.* 1992). Calculated ore reserves are shown in Table 2.

	A vein (Brocaigh)	C vein (Garrymore)
Grade	9.22g/t	10.39g/t
Tonnage (probable)	96,000t	152,000t
Tonnage (possible)	41,000t	209,000t

Table 2. Calculated ore reserves and grades for the Lecanvey gold prospect. g/t = grammes/metric ton.

Further south, close to Doo Lough in South Mayo, recent exploration has resulted in the discovery of significant gold mineralization, restricted to the Cregganbaun Shear Zone, mentioned in an earlier section. An excellent technical description of the deposits, and the exploration leading to their discovery, is provided by Thompson *et al.* (1992).

Gold mineralization has been recognised along the entire 33km length of the east - west trending Cregganbaun Shear Zone within the Sheeffry and Derrylea Formations. The shear zone lies chiefly within the Sheeffry Formation and is partially related to tuffaceous units within it (Thompson *et al.* 1992). Significant accumulations of gold have been identified at Tawnydoogan - Srahroosky (2406) and Cushinyen (2409).

At Tawnydoogan - Srahroosky, channel sampling across the outcropping part of the mineralized zone assayed 12g/t Au over 6m. Diamond drilling at this deposit has returned grades ranging from 19m at 2.1g/t Au to 2m at 28.7g/t Au. The mineralization occurs in sheared tuffs and other rocks over a strike length of 850m and to a depth of 160m. The width of the mineralized tuffs ranges from 1 to 14m and averages 7m. Gold occurs in its native form (Thompson *et al.* 1992). At Cushinyen, visible gold occurs in quartz pods in heavily deformed and altered sedimenatry and igneous rocks. The zone has been traced along strike for 250m and it is up to 10m wide. Trench channel samples returned grades of up to 19.6g/t Au over 2m and 15.3g/t Au over 7m. Assays from diamond drill samples ranged up to 10m at 9.1g/t Au. The gold occurs in quartz veins and is accompanied by minor amounts of pyrite, tetrahedrite (copper-antimony sulphide), sphalerite (zinc sulphide) and galena (lead sulphide). The host rocks are strongly altered, as marked by the abundance of carbonate, **sericite** and pyrite.

Gold mineralization has also been discovered on the island of Inishturk (5511) by Navan Resources plc. Visible gold mineralization occurs in the Middle Ordovician Derrylea Formation. Three east southeast trending main veins have been identified and termed the A,B and C veins. The veins are, on average, 1m to 2m thick and chiefly comprise quartz with abundant iron staining. The A vein has been traced for approximately 65m and has returned assays ranging from of 4.3g/t to 31.6g/t Au. The B vein has been traced for about 20m and has returned values of 2.1g/t to 2.3 g/t Au. The C vein is some 500m to 600m from the A vein and it has yielded a single value of 450ppb Au. Gold has also been discovered by Navan Resources at other localities on the Island.

Ovoca Gold Plc discovered a 1.6km long north-south trending breccia zone in the townland of Bohaun just south of Lough Kilbride. The zone consists of silicified breccias, with clasts of Silurian greywackes, from the Lettergesh Formation, embedded in massive quartz and quartz vein material. It displays evidence of polyphase deformation. Visible gold has been recorded and systematic sampling has returned gold values up to 420g/t. A drill hole sited to test the extension of the surface mineralization at depth intersected 8.7g/t Au over 0.8m at a depth of 36.4m. Gold grades tend to increase in shear zones within the breccia. These shear zones are sub-parallel to the margins of the breccia body.

LEAD - ZINC - COPPER DEPOSITS

The Sheet 10 area contains two lead vein deposits which were developed for exploratory or commercial purposes in the late 19th or early 20th Centuries, the Carrowgarriff or Clement's Mine, northwest of Lough Corrib and the Sheeffry Lead Mines (1271) in the Sheeffry Hills.

The Carrowgarriff or Clement's Mine, northwest of Lough Corrib, was developed by Clements Lead Mines Limited in 1907. Records suggest that the mine operated for only one year, although some of the dumps were scavenged in 1918. Crushing, dressing and washing plant were driven by water from a specially constructed reservoir on a small stream nearby.

The mine workings were on the surface and extended along a shear zone. A trench 150m long by 2.2m to 5m wide and up to 7m deep is all that remains of those workings today. An **adit** was driven beneath the surface workings, and connected to them by a **raise** and **drift**, in an attempt to locate additional mineralization, but without success. The adit was recently mapped on behalf of Aquitaine Mining (Ireland) Limited by D. Coller (GSI Open File).

The ore consists of a dense mass ("massive") of sulphides which apparently replace the marble host rock. Assays of dump material made in 1917 are given in Table 3.

Sample	Pb	Zn	Cu	Ag
A1	39.13%	5.38%	0.28%	167g/t
B2	47.04%	0.54%	0.65%	225g/t

Table 3. Assays of samples of dump material from the Clement's Mine.

The mine was worked primarily for its lead and silver content although it does contain iron, copper and zinc in their sulphide form (pyrite, chalcopyrite and sphalerite, respectively).

A deposit that has been termed the Sheeffry Lead Mines (1271) occurs in the Derrylea Formation within the Sheeffry Hills. A note by Cunningham (1950) describes in some detail the workings at this locality. These consist of five adit levels, apparently driven for exploratory, rather than production purposes. These trials date from the last century and there is little or no existing evidence of surface facilities. The greenish grey grits, which host the mineralization, dip to the north at 70° - 90°, and are cut out by faults and shear zones. One of the northeast-southwest trending shear zones is mineralized. The spoil heaps contain mineralized fault breccia with chalcopyrite and, despite the name of the deposit, subordinate galena.

COPPER - MOLYBDENUM DEPOSITS

Copper-molybdenum deposits are primarily associated with the Murvey Granite at Murvey and the Mace Head area at the western end of the Galway Granite and east southeast of Slyne Head. A detailed description of these deposits has been given by Max and Talbot (1986) upon which the following abbreviated account is based.

At Murvey (2196 and 5417) a deposit of 240,000t grading 0.13% Mo has been delineated. The deposit is some 270m long by 16m to 65m wide and extends in depth from 2.5m to 18m. The mineralization consists almost entirely of molybdenite (molybdenum sulphide) although some chalcopyrite is also present. Scattered occurrences of molybdenite occur within the granite up to 500m from the deposit, and also in the country rocks up to 70m from the granite contact. The molybdenite occurs in veins in a matrix of quartz and altered rock

The Mace Head (5398) deposit is smaller and no tonnage has been calculated. Molybdenite, chalcopyrite and pyrite occur disseminated within the **granodiorite** and molybdenite may also occur concentrated in syn- to late deformation quartz veins.

The Murvey and Mace deposits are situated in different positions within the Carna Dome of the Galway Granite. The Murvey deposit occurs at the margin of the Dome while the Mace deposit is more central and hence at a deeper level within the Dome. Other minor molybdenite, galena and sphalerite deposits occur near Rossaveel (2171), within an eastern sheet of the Murvey Granite.

IRON - COPPER - MOLYBDENUM - TUNGSTEN DEPOSITS

Layer parallel, iron sulphide rich lenses and disseminations, with or without minor chalcopyrite, molybdenite, scheelite (calcium-tungsten oxide) and wolframite (iron-magnesium-tungsten oxide), are one of two principal categories of mineral deposits distinguished by Reynolds *et al.* (1990) in the Dalradian rocks of Connemara. The lenses are generally restricted to certain stratigraphic units, most of the deposits occurring in the Lakes Marble Formation and a few occurring in the Ballynakill Schist Formation. The former formation consists of marbles, metavolcanics, schists and grits while the latter consists of aluminous schists, quartzites and pebble beds. It has also been suggested that the sulphide and tungsten mineralization is associated with the principal episode of volcanism during the deposition of the Lakes Marble Formation (C.B. Long, *personal communication).*

The deposits shown on the map, and described in further detail below, all occur at the eastern end of the Dalradian outcrop, close to the Oughterard Granite and the Maam Valley fault zone. Within the map area, the Derreennagusfoor-2, Drumsnauv-1 and Teernakill South-1 deposits occur within the Lakes Marble Formation and are in close proximity to the Oughterard Granite, while the Maumeen-2 deposit occurs within the Lakes Marble Formation but at a distance from the granite. The descriptions which follow are summarised from the descriptions of Reynolds *et al.* (1990).

Derreennagusfoor-2 (367): This steep dipping, lens shaped, iron-copper-tungsten body occurs northeast of Maam Cross. A surface excavation, now flooded, extends for 80m and is 3m to 8m wide. A 16m deep shaft was constructed to assist mining. Lakes Marble Formation amphibolites form both the footwall (rocks below) and hanging wall (rocks above) to the sulphide body which contains up to 80% sulphides. Pyrrhotite is the principal sulphide, in places partly replaced by pyrite, with minor chalcopyrite. There are no records of production from the deposit.

Drumsnauv-1 (373): Some mining trials were undertaken on this iron-copper-molybdenum deposit at some time in the past, but it is not known whether or not there was any production. A 3.5m shaft is recorded to have been sunk at the site, northwest of Lough Corrib, but there is little or no evidence of this today. Pyrite is the main sulphide present with minor chalcopyrite and molybdenite. Two veins which parallel the strike of the host Lakes Marble Formation, contain pyrite and chalcopyrite.

Maumeen-2 (2240): Substantial trials were made on pyritic mineralization at Maumeen, at the southeastern end of the Maumturk Mountains. A trench 30m long by 5m wide was dug to test the iron-copper-tungsten mineralization. Spoil heaps contain finely banded (1mm to 5mm) globular textured sulphides but elsewhere, parallel laminations are well developed. Sulphides make up to 80% of the rock with pyrite the principal sulphide. Minor pyrrhotite and chalcopyrite with some wolframite and scheelite are also present.

Teernakill South-1 (366): A mine, known as Teernakill or Northwest Mines, occurs at this site, northwest of Lough Corrib. Haughton (1860) recorded nickeliferous pyrrhotite from this iron-copper-molybdenum-tungsten deposit, but it was primarily worked as a sulphur and copper mine. The extent of the workings appears to be 60m long by 5m wide to a depth of some 14m. Lakes Marble Formation amphibolites form both the footwall and hanging wall to the mineralization. Pyrrhotite is the principal sulphide with minor pyrite and chalcopyrite and locally molybdenite.

"SKARN" DEPOSITS

"Skarn" deposits are generated by metamorphism of carbonate rich rocks adjacent to igneous intrusions, resulting in the formation of a diverse range of sometimes quite unusual minerals, as well as metallic and other minerals of potential commercial interest. In the Sheet 10 area, a number of small skarn deposits are associated with the Omey Granite and in the small granitic bodies at the western end of Connemara. These were discovered during the course of mineral exploration work undertaken by the Anglo United/Irish Base Metals consortium in the 1970s and 1980s.

The skarns occur in calcareous country rocks, principally in the Lakes Marble Formation, and range up to 6m wide. A wide range of typical "skarn" minerals occur in the various deposits, which include one or more of the following mainly metallic sulphide minerals of interest: scheelite, galena, molybdenite, bismuthinite (bismuth sulphide), chalcopyrite, fluorite (calcium fluoride), pyrrhotite and the semi-precious gem stones, beryl and garnet. Values of up to 0.5% tungsten (WO_3) have been recorded from grab samples.

Eight principal skarns have been identified, the mineralogy of which is summarized in Table 4.

It is possible that the Lakes Marble Formation was already enriched in tungsten and that this, and other metals, were redistributed and concentrated into skarns by intrusion of the Galway Granite.

Skarn	Mineralogy
Sellarna (5515)	Scheelite, magnetite, galena, molybdenite, some bismuthinite (0.75% Bi)
Courhoorlough (5516)	Scheelite, molybdenite
South Courhoor (5517)	Galena, molybdenite, scheelite, chalcopyrite, garnet
Glen (5518)	Pyrrhotite, scheelite and garnet
Cushatrough (5519)	Pyrrhotite, scheelite and garnet
Barnahallia Lough (5520)	Scheelite 2% tungsten oxide, (WO_3) and garnet
Kingstown Peninsula (5521)	Pyrrhotite (up to 25%), chalcopyrite, scheelite (0.2% WO_3)
Leagaun (5522)	Garnet (50%),

Table 4. The mineralogy of the principal skarn deposits associated with the Omey Granite.

COPPER - NICKEL PLATINUM - PALLADIUM AND CHROME DEPOSITS

In the Cashel-Lough Wheelaun District, layered mafic and **ultramafic** intrusive rocks contain Cu-Ni mineralization within **pyroxenites** at the base of the intrusion. Minor platinum and palladium values have been recorded from the intrusion. Chromite layers (chromium oxide) occur in the Dawros **peridotite** (358), west of Kylemore Abbey. Reynolds *et al.* (1990) note that the layers may range up to 10cm thick, although thinner 2.5cm layers are more common.

IRON DEPOSITS

A 6-20m thick stratabound magnetite (iron oxide) layer occurs in southwest Gorumna Island, on Lettermullan Island (5409 and 5410) and on Golam (5405) Island. This layer has been traced for over a kilometre along strike and may represent an originally continuous horizon, now disrupted. It contains up to 20% magnetite, based upon visual estimate. The body appears to be associated with an amphibolite intrusion which itself has been intruded between the Ryans Farm and Golam Formations.

BARYTE DEPOSITS

Some barytes veins occur within the Cregganbaun Formation just to the south of Westport. Two small prospects were discovered by Tara Prospecting Limited at Knappaghbeg and Lanmore. Barytes occurs as open space filling in fault zones near the Cregganbaun basal unconformity (Tara Prospecting Limited reports, GSI Open File). Some base metal (lead and zinc) mineralization occurs in association with the barytes.

Industrial Minerals and Rocks

DIMENSION STONE

"Dimension stone" is the name given to any type of rock used for building, roofing, monumental or decorative purposes of any type.

Granite

A number of distinctive dimension stone types occur within the Galway Granite. Colours vary from pink to grey and grey with pink feldspar phenocrysts. There are no currently working dimension stone quarries listed for the Galway Granite (Claringbold *et al.* 1994), although there are a significant number of localities that have been classified as having potential for dimension stone production (Bell 1992). These include a pink granite from the Omey Granite at Aughrus More (3983); a medium grained grey granite from within the Carna Granite at Cuilleen (3990); two grey granites from the Spiddal Granite; Derrykyle (4020) and Cornarona (4022); and two grey granites from the Errisbeg Townland Granite from Glendruid (3997) and Formoyle (4017).

Marble

The marbles of Connemara are typically green serpentinous marbles, although they vary in colour from green and white to brown and grey (Max 1985). The marbles are used mainly for ornamental purposes, though due to susceptibility to weathering and graphite content which leaves a grey residue on weathered surfaces, they are not ideally suited to external applications.

Currently there are two marble quarries operating in Co. Galway (Claringbold *et al.* 1994). These are located in the townlands of Lissoughter (2277) and Barnanoraun (418). Lissoughter Quarry is historically important as a source of Connemara marble and has been used in such buildings as the Museum Building at Trinity College, Dublin. The quarry is currently producing 1m cubic blocks that are processed for interior tiles and giftware. Barnanoraun Quarry is a smaller quarry producing paler green Connemara marble in a range of block sizes. There are also quarries that are sporadically active, depending on demand. Streamstown quarry (416: see back cover plate) is noted for the fine figuring of the marble.

Quartzite

Iron and manganese stained, pale grey quartzite is currently being quarried for dimension stone at Kildownet (GR 07125 29545) in the extreme southeast of Achill Island. The quartzite forms part of the Kildownet Quartzite Member of the Dalradian Cove Schist Formation. Bed thickness in the quartzite influences specific applications, which includes use as cladding, paving or as small blocks.

Serpentinite

There are a significant number of serpentinite localities in the area, especially around Clew Bay. None of the occurrences are currently in production, although there are a number of sites that are believed to have potential, such as Falduff (1110) and Lenacraigaboy (1108). Serpentinite may be used as a dimension stone, as decorative aggregate or as an ornamental stone for carved or turned articles.

Slate
There are twelve recorded slate quarries all of which are now disused. These quarries are typically small and would have provided a local source of roofing slate, mostly of poor quality.

AGGREGATES
"Aggregates" is the term applied to any rock or mineral used in bulk, in crushed or ground form, for road or building construction or materials fabrication, or for agricultural, agri-chemical or general chemical applications.

Sand and Gravel
Glacial and fluvial deposits in this area provide a natural source of aggregate for local use. The deposits vary from very fine sands, sands and gravels through to large cobbles and boulders. There is one pit currently operating in Addergoole (5278) which is supplying sand and gravel suitable for concrete in the sizes 160mm, 80mm, 15mm and graded and washed sand. There are approximately another twenty sand and gravel locations in this area, some of which are sporadically active, depending on demand.

Granite
There are no operating aggregate quarries in the Galway Granite although in the past these granites have been used as a source of aggregate. In contrast to dimension stone extraction, granites with a closer joint spacing are more suitable for aggregate production.

Metasedimentary rocks
Metasedimentary rocks, chiefly psammites and quartzites, have been quarried for aggregate in the past and are currently being extracted from an active quarry just west of Recess Post Office. This quarry occurs in strongly banded calcareous psammites, part of the Upper Banded Member of the Streamstown Schist Formation. Coarser metasedimentary rocks may have potential as road surfacing aggregates, since they give good skid resistance.

OTHER MINERALS

Talc
There are a number of talc occurrences in this area, the most significant of which occur on Inishbofin Island and at Corrig Hill near Westport.

The Inishbofin deposit (405) is an alteration product derived from serpentinized ultramafic rocks adjacent to the Renvyle-Bofin Slide (Cruse and Leake 1968). It consists of approximately 1 million tonnes of talc-magnesite rock, composed of about 50% talc, 10% platy serpentinite and 40% carbonates (Bell and Flegg 1979). This deposit has not been commercially exploited due to lack of infrastructure on the island and the likely environmental impact of quarrying on the tourism and fishing industries of the island.

A large talc deposit occurs at Corrig Hill, near Westport. Exploration work indicates that there is approximately 3.5 million tonnes of talc-magnesite in the townland of Killaghoor (1103) and surrounding area. Talc-magnesite makes up 75%-95% of the rock, the remainder consisting of serpentine and other minerals. The rock is fibre and silica free (Hough 1990). Planning permission to develop this deposit was not granted on the grounds that the effects of quarrying would damage the landscape in such a scenic area and thus affect the tourism industry (GSI I.M. Newsletter 1991).

Fluorite
Fluorite mineralization is associated with metallic mineral localities in the Dalradian and also with mineralization in the Galway Granite. In Cartron townland (5236), fluorite mineralization occurs in greisenized and garnetiferous Murvey Granite. O'Connor *et al.* (1993) suggest that the hydrothermal fluids responsible for ore deposition were probably generated in mid-Triassic times in response to continental rifting of the Atlantic margin of western Ireland.

Quartz
The Bennabeola Quartzite Formation contains a number of large quartz veins and lodes. One of these, on Bengower (414), is described by Bishopp and McCluskey (1948). Irish Quartz Ltd have prospected in Derrynauglan townland (41) where there is a quartz vein 10m by 50m by 2500m and in Derryclare townland (42) where there is a 450m long quartz vein. Quartz vein mineralization is also associated with the Galway Granite. At Carrowroe (5360) a feldspathic vein may be traced for about three miles with thicknesses of 15m expanding to 100m at one point (Bishopp and McCluskey 1948).

Asbestos
There are numerous small asbestos deposits in this area, most of which are associated with the Dawros peridotite intrusion, for example at Dawros More (5275). These deposits, as well as the serpentinites

along the south shore of Clew Bay, have been investigated as potential sources of commercial asbestos, although none has been mined.

GROUNDWATER RESOURCES

Introduction

Groundwater is the water that is stored in and moves through the pores and cracks in subsoils and bedrock. Overall it provides about 25% of the drinking water supplies in Ireland, although this proportion varies from place to place, depending on the underlying geology and the water needs, from a few percent to over 90%. The potential of rocks to yield water in significant quantities depends on the permeability (permeability is a measure of the capacity of a rock to transmit water). There are two types of permeability - intergranular and fissure. Intergranular permeability is found in unlithified sediments such as sands, gravels, tills, clays and peats, where groundwater movement is between the grains. Fissure permeability occurs in bedrock where water movement is through cracks, joints, bedding planes and solution openings. The permeability of bedrock can vary by several orders of magnitude and, among the factors which this variation depends on, are lithology or rock type and structural history, in particular the degree of fracturing.

Aquifers are rocks that contain sufficient voids to store water and are permeable enough to allow water to flow through them in significant quantities. Rocks capable of yielding large water supplies (>400m^3/day) from wells are classed as "regionally important" aquifers; rocks capable of giving sufficient water for small group schemes or villages (100-400m^3/d) are classed as "locally important"; with low permeability rocks classed as "poor" aquifers.

The remainder of this brief account considers the aquifer potential and the permeability of the rocks in the area covered by Sheet 10. Few hydrogeological investigations have been carried out in the area, consequently the content of this account is based on general hydrogeological principles and information from other areas.

Basement Rocks

The granites, together with the Lower Palaeozoic and Precambrian rocks that underlie most of Sheet 10, have similar hydrogeological characteristics and therefore can be considered together. They are characterised by a low fissure permeability. Consequently they are classed as poor aquifers which are generally unproductive. Even so, enough water may usually be obtained from bored wells for domestic supplies (10-20m^3/d), although some failed wells can be expected. Most groundwater in these areas moves in the upper fractured zone, more permeable beds of limited extent and along fault or fracture zones. The flow is generally in localised systems with little continuity between them. The low storage in these strata is usually balanced by the higher rainfall on uplands. However, during long dry spells, baseflow to streams can be reduced significantly as many of the springs and seepages which feed them dry up. Well yields are greatest in the fault zones, particularly in the more coarse grained rocks such as the quartzites, although even in these areas yields greater than 100m^3/d would be exceptional. Yields are lowest in the fine grained metamorphic rocks. **Felsite** and **pegmatite** veins in the granite are usually more permeable than the surrounding rocks. At a factory site in Spiddal, three boreholes in the pegmatite gave substantial yields; a fourth in the granite gave no water.

Old Red Sandstone

Evidence from County Roscommon suggests that the Devonian sandstones and conglomerates have a low permeability and limited groundwater potential, with boreholes usually capable of yielding no more than domestic supplies. Higher yields are likely along fault zones, but even here yields greater than 200m^3/d are unlikely. These rocks are classed as poor aquifers.

Carboniferous Sandstone

These sandstones are likely to have a higher permeability than underlying Devonian red sandstones and conglomerates and yields of 100-400m^3/d may be obtained. Highest yields are likely in fault zones. They are classed as locally important aquifers.

Carboniferous Limestones

The development of permeability in the limestones depends largely on the clay content - generally the purer and cleaner the limestone, the higher the permeability. Consequently the Barney and Westport Limestones are likely to have the greatest aquifer potential and can be classed as regionally important aquifers. Solution of these limestones by percolating water is likely and so they are probably **karst**ified to a significant degree. The other limestones have a higher clay and shale content and are classed as locally important aquifers although borehole yields are likely to vary, with highest yields found in the purer and more fissured areas.

Quaternary Sediments

The hydrogeological significance of these sediments is very variable and is largely a function of their permeability, thickness and extent. The low permeability material (till, clay and peat) protects underlying groundwater, restricts recharge and, where sufficiently thick, may confine the groundwater. The high permeability materials (sands and gravels) allow a high level of recharge, provide additional storage to that of the underlying bedrock aquifers and where sufficiently thick, can be an aquifer in their own right.

As sands/gravels are not common in the area, there are no regionally important sand/gravel aquifers. Locally important water supplies could probably be found in the Letterfrack/Kylemore area and in any other areas where the gravels are present in low lying situations.

Hydrochemistry and Water Quality

Groundwaters in the basement rocks and the Devonian conglomerates are relatively soft with a total hardness less than 150 mg/l (as $CaCO_3$) and often less than 100 mg/l. Harder groundwaters are found where they are overlain by Quaternary deposits that contain limestone or in areas underlain by limestones. In such areas, high iron and manganese concentrations are a common problem. The limestones contain mainly calcium bicarbonate water types. Total hardness in these rocks generally ranges from 250 mg/l (as $CaCO_3$) to over 400 mg/l.

The groundwater in these strata is normally of potable quality except for small local areas where they have been contaminated, mainly by the effluent from organic wastes (e.g. farmyards and septic tank systems).

Conclusion

Of all the 1:100,000 sheets covering Ireland, the area of Sheet 10 is among the worst from the point of view of groundwater potential. Most of the rocks have a low permeability and poor aquifer potential. Groundwater is not a major source of drinking water for public supplies, although many private wells are in use.

BIBLIOGRAPHY

The following list of general, introductory texts is provided in order to assist any interested reader to delve a little further into the subjects of geology, landscape and the Earth Sciences in general.

DUNNING, F.W., ADAMS, P.J., THACKRAY, J.C., van ROSE, S., MERCER, I.F., and ROBERTS, R.H. 1981. *The story of the Earth.* (2nd Edition). Geological Museum, Institute of Geological Sciences, London, H.M.S.O., 36pp.

DUNNING, F.W., MERCER, I.F., OWEN, M.P., ROBERTS, R.H. and LAMBERT, J.L.M. 1978. *Britain before Man.* Geological Museum, Institute of Geological Sciences, London, H.M.S.O., 36pp.

HOLMES, A. 1978. *Principles of Physical Geology* (3rd edition revised by D.Holmes). Thomas Nelson, London and Edinburgh, 730pp.

MITCHELL, F. 1986. *The Shell Guide to Reading the Irish Landscape.* Michael Joseph/Country House, 228pp.

THACKRAY, J. 1980. *The age of the Earth.* Geological Museum, Institute of Geological Sciences, London, H.M.S.O., 36pp.

WHITTOW, J.B. 1974. *Geology and Scenery in Ireland.* Pelican Books, 301pp.

For more detailed and technical consideration of the evolution of the geology of Ireland and Britain, and the Connemara/South Mayo region, the following texts are suggested.

ANDERTON, R., BRIDGES, P.H., LEEDER, M.R. and SELLWOOD, B.W. 1979. *A Dynamic Stratigraphy of the British Isles.* George Allen and Unwin Ltd, 301pp.

GRAHAM, J.R., LEAKE, B.E. and RYAN, P.D. 1985. *The geology of South Mayo;* a 1 inch to 1 mile colour printed compilation map. University of Glasgow.

GRAHAM, J.R., LEAKE, B.E. and RYAN, P.D. 1989. *The Geology of South Mayo.* Scottish Academic Press.

HOLLAND, C.H. (editor) 1981. *A Geology of Ireland.* Scottish Academic Press, 335pp.

LEAKE, B.E. and TANNER, P.W.G. 1994. *The geology of the Dalradian and associated rocks of Connemara, western Ireland.* Royal Irish Academy, 96pp.

LEAKE, B.E., TANNER, P.W.G. and SENIOR, A. 1981. *The Geology of Connemara.* 1:63,360 (1 inch to 1 mile) scale colour printed compilation map, University of Glasgow.

LEAKE, B.E., TANNER, P.W.G. and SENIOR, A. 1981. *Fold Traces (F3 to F5) and Regional Metamorphism in the Dalradian Rocks of Connemara.* 1:63,360 (1 inch to 1 mile) scale colour printed compilation map, University of Glasgow.

LONG, C.B. 1992. *Brief lithological descriptions of all Cambro-Ordovician, Dalradian, Pre-Dalradian and igneous rock units of Sheet 6, North Mayo.* A supplement to accompany "Geology of North Mayo". Geological Survey of Ireland. 18pp.

LONG, C.B., McCONNELL, B., ARCHER, J.B and others. 1995. *The Geology of Connemara and South Mayo.* Geological Survey of Ireland.

MAX, M.D., LONG, C.B. and MacDERMOT, C.V. 1992. *Bedrock Geology of North Mayo.* Geological Survey of Ireland, Sheet 6, 1:100,000 scale colour printed map.

TANNER, P.W.G. 1981. *Cross sections for The Geological Map of Connemara.* 1:63,360 (1 inch to 1 mile) scale. University of Glasgow.

REFERENCES CITED IN THE TEXT

AHERNE, S., REYNOLDS, N.A. and BURKE, D.J. 1992. Gold mineralization in the Silurian and Ordovician of South Mayo. *In*: Bowden, A.A., Earls, G., O'Connor, P.G. and Pyne, J.F. (eds), *The Irish Minerals Industry 1980-1990.* Irish Association for Economic Geology, 39-49.

BELL, A. 1992. The Galway granites: dimension stone potential. *Geological Survey of Ireland Report Series* RS 92/1, 113p

BELL, A. and FLEGG, A.M. 1979. The talc magnesite deposits of Inishbofin and Inishshark (Detailed petrological analyses and results.) *Geological Survey of Ireland Unpublished Report*, 42pp

BISHOPP, D.W. and McCLUSKEY B.E. 1948. Sources of industrial silica in Ireland. *Geological Survey of Ireland Emergency period Pamphlet* No.3. 5p.

BOSENCE, D.W.J. 1979. Live and dead faunas from coralline algal gravels, Co. Galway, Ireland. *Palaeontology* 22, 449-478.

CLARINGBOLD, K., FLEGG, A., MAGEE, R. and VONHOF, J. 1994. Directory of active quarries, pits, and mines in Ireland. *Geological Survey of Ireland Report Series* RS 94/4, 111p.

CRUSE, M.J.B. and LEAKE, B.E. 1968. The geology of Renvyle, Inishbofin and Inishshark, North-west Connemara, Co. Galway, Eire. *Proceedings of the Royal Irish Academy,* 67B, 1-36.

CUNNINGHAM, M.A. 1950. The Sheeffry Mine. *Internal Geological Survey of Ireland Report*, 3pp.

DALZIEL, I.W.D. 1995. Earth before Pangea. *Scientific American,* January 1995, 58-63.

HAMILTON, W.R, WOOLLEY, A.R. and BISHOP, A.C. 1981. *The Hamlyn Guide to Minerals, Rocks and Fossils.* Hamlyn Publishing Group Ltd, London, 320pp.

HAMMOND, R. F. 1981. The Peatlands of Ireland (2nd ed.). *Soil Survey Bulletin No 35, An Foras Talúntais,* Dublin, 60pp.

HARLEY, M.J., NIELD, T. and McKIRDY, A.P. 1990. *Death of an Ocean.* UK Nature Conservancy Council, Peterborough, 16pp.

HAUGHTON, S. 1860. On the occurrence of nickeliferous magnetic pyrites from Tiernakill, near Maum, County Galway. *Journal of the Geological Society of Dublin,* 9, 1-2.

HOUGH, D. 1990. Ivernia West acquires Talc Technology. *Geological Survey of Ireland Industrial Minerals Newsletter* No. 11, p2.

KENNAN, P. 1995. *Written in Stone.* Geological Survey of Ireland, 50pp.

LONG, C.B., MacDERMOT, C.V., MORRIS, J.H., SLEEMAN, A.G., TIETZSCH-TYLER, D., ALDWELL, C.R., DALY, D., FLEGG, A.M., McARDLE, P.M. and WARREN, W.P. 1992. *Geology of North Mayo.* A geological description to accompany the bedrock geology 1:100,000 map series; sheet 6, North Mayo. Geological Survey of Ireland Map Report Series, 50pp.

MAX, M.D. 1985. Connemara Marble and the Industry based upon it. *Geological Survey of Ireland Report Series,* RS 85/2, 32pp.

MAX, M.D. and TALBOT, V. 1986. Molybdenum concentrations in the western end of the Galway Granite and their structural setting. *In*: Andrew, C.J., Crowe, R.A., Finlay, S., Pennell, W.M. and Pyne, J.F. (eds), *Geology and Genesis of Mineral Deposits in Ireland.* Irish Association for Economic Geology, 177-185.

MOHR, P. 1982. Tertiary dolerite intrusions of west-central Ireland. *Proceedings of the Royal Irish Academy,* 82B, 53-82.

MOHR, P. 1986. Possible Late Pleistocene faulting in Iar (West) Connacht, Ireland. *Geological Magazine,* 123, 545-552.

MOHR, P. 1993. Composite dolerite-granophyre dikes and associated granitic intrusions of Carboniferous age in eastern Connemara, Ireland. *Irish Journal of Earth Sciences,* 12, 119-138.

NIELD, T., McKIRDY, A.P. and HARLEY, M.J. 1989. *The age of Ice.* UK Nature Conservancy Council, Peterborough, 16pp.

O'CONNOR, P.J., HOGELSBERGER, H., FEELY, M., and REX, D.C. 1993. Fluid inclusion studies, rare earth element chemistry and age of hydrothermal fluorite mineralization in Western Ireland - a link with continental rifting? *Transactions of the Institute of Mining and Metallurgy,* (Sect. B: Applied Earth Sciences), 102, B141-B148.

REYNOLDS, N., McARDLE, P., PYNE, J.F., FARRELL, L.P.C. and FLEGG, A.M. 1990. Mineral localities in the Dalradian and associated igneous rocks of Connemara, County Galway. *Geological Survey of Ireland Report Series,* RS 90/2, 89pp plus map.

THOMPSON, S.J., SHINE, C.H., COOPER, C., HALLS, C. and ZHAO, R. 1992. Shear-hosted gold mineralization in Co. Mayo, Ireland. *In*: Bowden, A.A., Earls, G., O'Connor, P.G. and Pyne, J.F. (eds), *The Irish Minerals Industry 1980-1990.* Irish Association for Economic Geology. 21-37.

WARREN, W. P. 1992. Drumlin orientation and the pattern of glaciation in Ireland. *In*: Robertsson, A-M., Ringberg, B., Miller, U and Brunnberg, L. (eds), *Quaternary Stratigraphy, Glacial Morphology and Environmental Changes.* Research Papers, series Ca 81. Sveriges Geologiska Undersökning, Uppsala, 359-366.

APPENDICES

APPENDIX 1: SIMPLIFIED DESCRIPTIONS OF ROCK UNITS

This appendix provides brief lithological descriptions for all stratigraphic units appearing on Sheet 10, the Geology of Connemara. The units are listed in alphabetical order defined by the unit codes used on the map and in the booklet.

The first rock type listed in *italics* may be used to characterise the entire unit at a very basic level, although the reader should be aware that, in most instances, each rock unit consists of many specific rock types. The prefix "*meta-*" is added, where appropriate, to indicate that the name refers to a metamorphosed equivalent of the original rock type.

AA Aille and Barney Limestone Formations (undifferentiated). *Limestone.*
Grey limestones and minor shales, with clean, pale grey, thick bedded limestones towards the top.

AAwp Westport Oolite Member. *Limestone and sandstone.*
Cross bedded, sandy oolitic limestones interbedded with subordinate mudstone and calcareous sandstone.

AB South Achillbeg Formation. *Greywacke - slate.*
A mixed sequence of grey, green and black slates, greywackes, quartzites and conglomerates.

ABcg Achillbeg Conglomerate Member. *Grit and Slate.*
Pebbly sandstones, conglomerates and slates.

ABps Achillbeg Lighthouse Psammite Member. *Slate and quartzite.*
Grey and green slates and calcareous quartzites.

ABsl Achillbeg School Black Slate Member. *Slate.*
Black slates with psammites and pale quartzites.

AD Ashleam Bridge Dolomitic Formation. *Marble.*
Buff to white dolomitic marble and dolomitic and graphitic schists.

AI Aille Limestone Formation. *Muddy limestone.*
Dark grey, fine grained limestone and minor thin shales.

AN Anaffrin Formation. *Coarse schist.*
Mainly psammitic schists, with local quartzitic psammites.

ANgm Glennamong Member. *Schist.*
Pelitic and semi-pelitic schists.

AP Ashleam Head Formation. *Quartzite.*
Pale green quartzites, with minor schists and chloritic metatuffs.

APsh Ashleam Schist Member. *Schist.*
Calcareous psammitic and semi-pelitic schists, with minor quartzite.

AQ Ashleam Bridge Quartzite Formation. *Mixed quartzite - schist.*
Pale pebbly quartzites with black pelitic schists.

AS Ashleam Bay Formation. *Schist.*
Black graphitic pelitic schists with dolomitic marbles and pale quartzites.

AT Atlantic Drive Schist Formation. *Mixed quartzite - schist.*
Brecciated and finely bedded schistose quartzites with schists.

BA Ballytoohy Formation. *Mudstone with blocks of all sizes.*
Black shale matrix with blocks of chert and sandstone. Mélange.

BaAm Ballyconneely Amphibolite. *Meta-basalt.*
Intensely banded, fine grained meta-basaltic rocks; mylonite.

BC Bencorragh Formation. *Basalt.*
Green, basalt lavas, commonly vesicular and pillowed.

BH Birreencorragh Schist Formation. *Schist.*
Grey graphitic schists and minor grey quartzites.

BHqz Birreencorragh Quartzite Member. *Quartzite*
Quartzites and gritty quartzites with minor schists.

BI Birreen Formation. *Conglomerate.*
Conglomerates composed of clasts of various igneous rocks and red sandstone.

BK Ballynakill Schist Formation. *Schist.*
Brown to grey pelitic schists with psammitic schists and quartzites, locally gritty or pebbly.

BN Bohaun Volcanic Formation. *Basalt.*
Mainly pillowed and vesicular basalt lavas with interlayered chert and tuff.

BO Buckoogh Formation. *Schist.*
Mixed schists with minor pebbly grits.

BR Bills Rocks Formation. *Greywacke.*
Grey and green coarse grained greywackes and conglomerates with siltstone and mudstone.

BS Bunnamohaun Siltstone Formation. *Siltstone.*
Red and green siltstone, minor sandstone.

BU Bouris Formation. *Mixed schist - limestone.*
Mixed pelitic and psammitic schists, limestones and minor tuffs.

BV Bunaveela Lough Formation. *Schist.*
Mixed schists with minor basic metavolcanics and marble.

BW Ben Levy Grit Formation. *Schist.*
Grey-green pelitic to psammitic schists, thin quartzites, locally pebbly, and various basic volcanic and intrusive rocks. Mylonitic.

BX Bennabeola Quartzite Formation. *Quartzite.*
Mainly white, pale green or grey massive or bedded, orange to pink weathering quartzites, locally gritty or pebbly.

BY Bunnafahy Conglomerate Formation. *Quartzite.*
White quartzite, psammite and conglomerate.

BZ Barnanoraun Schist Formation. *Schist.*
Mixed brown and grey schists, sometimes calcareous, and pale green and brown basic gneisses.

CB Cregganbaun Formation. *Quartzite.*
Mixed quartzites, psammites, conglomerates and semi-pelitic schist.

CC Cleggan Boulder Bed Formation. *Meta-conglomerate.*
Coarse pebble/cobble bearing psammites and semi-pelitic schists of glacial-marine origin.

CE Cove Schist Formation. *Mixed schist - quartzite.*
Black pelitic schists and thin pale to black quartzites.

CJ Connemara Marble Formation. *Marble.*
Dolomitic green, brown, grey and white marbles and calcareous schists.

CL Cashel Schist Formation. *Schist.*
Mixed psammitic and pelitic schists, locally pebbly, gneisses common in southern part of unit.

CM Croaghmoyle Formation. *Conglomerate.*
Conglomerates, composed mostly of quartzite clasts.

CN Corraun Formation. *Mixed quartzite - schist.*
A mixed sequence of quartzites and schists divided into 4 members.

CNas Corraun Schist Member (Lower). *Schist.*
Dominantly black pelitic and semi-pelitic schists, with other minor types including chloritic metavolcanic greenbeds.

CNbs Corraun Schist Member (Upper). *Schist.*
Dark green and black schists, with minor calcareous schists and tuffaceous metavolcanics.

CNlq Corraun Quartzite Member. *Quartzite.*
White and grey quartzites and banded psammitic schists.

CNuk Kildownet Quartzite Member. *Quartzite.*
Pale quartzites and minor tuffaceous basic metavolcanics.

CoGr Corvock Granite. *Granite.*
A named body of various types of granitic (Gr) rocks, including granodiorite.

CP Capnagower Formation. *Sandstone.*
Grey sandstone and siltstone.

CR Castlebar River Limestone Formation. *Muddy limestone.*
Dark, fine grained limestone and calcareous shale, minor oolitic limestone at base.

CS Cullydoo Formation. *Quartzite.*
Mainly thin banded, white quartzite and psammitic schist.

CSsh Srahmore Quartzite and Schist Member. *Mixed quartzite - schist.*
Thinly banded white quartzites and semi-pelitic schists.

CSsq Srahmore Quartzite Member. *Quartzite.*
Banded white quartzites and pale psammitic schists.

CV Claggan Volcanic Formation. *Meta-volcanics.*
Metamorphosed mafic tuffs and tuffaceous semi-pelitic schists.

CW Cornamona Marble Formation. *Mixed schist - marble.*
Mixed black schists, marbles and marble breccias.

D Dolerite and Gabbro. *Basalt.*
Dark green fine (dolerite) and coarse (gabbro) grained basaltic rocks.

DA Dooega Head Formation. *Quartzite.*
Pale coloured quartzites and psammitic schists.

DB Derry Bay Formation. *Mixed tuff - shale.*
Mixed sequence of fine to coarse volcanic tuffs and minor black shales and siltstones.

DE Derreen Marble Formation. *Marble.*
Brown dolomitic marble with subordinate siliceous interbeds.

DEsh Derreen Schist Member. *Schist.*
Calcareous psammitic and minor semi-pelitic schists.

DF Delaney Dome Meta-rhyolite Formation. *Meta-granite.*
Intensely banded, pink to grey acid igneous rocks of granitic type. Mylonite.

DG Derryheagh Formation. *Marble - siltstone.*
Marbles and fossil bearing calcareous siltstones.

DIN Dinantian limestones (undifferentiated). *Limestone.*
Undifferentiated limestones of Carboniferous (Dinantian) age.

Di Diorite. *Diorite.*
Medium grained, mottled green-white igneous rocks, intermediate between basalt and granite.

DL Derrylea Formation. *Greywacke.*
Mixed interbedded greywacke and slate, some tuffs.

DM Derrymore Formation. *Greywacke.*
Mixed sequence of sandstones and mudstones, with some conglomerates and tuffs.

DP Deer Park Schist Formation. *Schist - meta*-volcanics
Semi-pelitic schists and metamorphosed basaltic volcanic rocks.

DV Derryveeny Formation. *Conglomerate.*
Conglomerates and sandstones.

DX Deer Park Complex. *Serpentinite.*
Metamorphosed basaltic and ultramafic ocean crust/mantle igneous rocks.

ER Extrusive Rhyolite Formation. *Rhyolite.*
Pinkish coloured, very fine grained granite (= rhyolite) lavas and breccias.

F Felsite. *Rhyolite.*
Creamy white, flinty textured, extremely fine grained intrusive granitic rocks.

FN Farnacht Formation. *Andesite.*
Metamorphosed dacitic to andesitic lavas.

FY Finny Formation. *Mixed tuff - chert.*
A mixed sequence of acid volcanic breccias, tuffs, lavas and cherts.

GaBz Banded (= Magma Mixing) Zone. *Banded Granite.*
A named, 4-6km wide zone of various intersheeted varieties of granite (Gr) in the east part of the Galway Granite (GaGr).

GaCf Callowfinish Granite. *Granite.*
A named body of the Galway Granite (GaGr): pink coloured granite (Gr), with large feldspar crystals.

GaCn Carna Granite - Carna type. *Granite.*
A named body of the Galway Granite (GaGr): medium grained, grey coloured granodiorite, with some large feldspar crystals.

GaCu Carna Granite - Cuilleen type. *Granite.*
A named body of the Galway Granite (GaGr): pink coloured granite with large pink feldspar crystals.

GaEb Errisbeg Townland Granite. *Granite.*
A named body of the Galway Granite (GaGr): pink to pale grey coloured granite, with large pink feldspar crystals.

GaGr GALWAY GRANITE. (undifferentiated). *Granite.*
A regional scale cluster of granite and granodiorite plutons.

GaLe Lettermore Granodiorites. *Granite.*
A named body of the Galway Granite (GaGr): a grey, medium grained granodiorite.

GaMa Mafic Errisbeg Townland Granite. *Granite.*
A dark variety of the Errisbeg Townland Granite (GaEb) with large feldspar crystals.

GaMu Murvey Granite. *Granite.*
A pink coloured body of granite, part of the Galway Granite (GaGr).

GaSh Shannawona Granite. *Granite.*
A named body of the Galway Granite (GaGr): similar to the Callowfinish Granite (GaCf) but darker in colour.

GaSp Spiddal Granite. *Granite.*
A named body of the Galway Granite (GaGr): medium grained, grey coloured granodiorite.

GC Glencraff Formation. *Sandstone.*
Thin bedded sandstones and mudstones.

GM Graffa More Formation. *Conglomerate.*
Red and green conglomerates, with minor breccias and coarse sandstones.

GO Golam Formation. *Chert.*
Red, green and black cherts and interbedded shales.

GP Glen Pebbly Arkose Formation. *Sandstone - conglomerate.*
Conglomerates, coarse sandstones and minor red mudstones.

GR Gorumna Formation. *Meta-basalt.*
Metamorphosed basalt lavas and intrusions, with lenses of sedimentary rocks (GRs).

GRs Sedimentary lenses (main occurrences). *Mixed meta-sediments.*
Lenses of mixed rock types within the Gorumna Formation (GR): Chert, pebbly mudstone, sandstone, tuffs and agglomerates.

Gr Minor Granitic Bodies. *Granite*
Light coloured, coarse to fine grained igneous intrusive rocks composed of quartz, feldspar and mica.

GU Glenummera Formation. *Slate.*
Slates with very minor siltstone and sandstone.

GV Glenlara Volcanic Formation. *Meta-volcanics.*
Basic metavolcanic tuffs and lavas.

H Metadolerite or amphibolite. *Meta-basalt.*
Intrusive bodies of fine to medium grained meta-basalt or meta-dolerite.

InGr Inish Granite. *Granite*
A named body of a pink coloured variety of granite.

IR Kinrovar Schist. *Schist.*
Pelitic and semi-pelitic, sometimes banded schists.

KB Kilbride Formation. *Sandstone.*
Dominantly green and grey sandstones, with subordinate mudstones, sometimes purple, conglomerates and tuffs.

KI Killadangan Formation. *Pebbly mudstone.*
Black shale containing large blocks of chert, sandstone and limestone.

KK Knock Kilbride Formation. *Basalt.*
Dark green basaltic lavas, sometimes pillowed and vesicular, with red and green cherts, breccias and shales.

KM Knockmore Sandstone Formation. *Sandstone.*
Green and buff sandstones with minor red mudstone.

KN Knockcorraun Formation. *Schist.*
Mainly black and grey graphitic and calcareous schists.

KNma Knockcorraun Lough Marble Member. *Marble.*
Dolomitic marble with minor schists.

KS Kill Sandstone Formation. *Sandstone.*
Sandstone and conglomerate.

LC Lough Kilbride Schist Formation. *Schist.*
Interbedded, intensely banded psammitic and pelitic schists, locally pebbly; mylonitic.

LeGr Letterfrack granitic plutons. *Granite.*
A group of grey coloured granodioritic bodies.

LF Loch Faoilean Formation. *Meta-basalt.*
Metamorphosed volcanic and intrusive basalts, with minor limestones and limestone breccias.

LG Lettergesh Formation. *Greywacke.*
Greywacke sandstones, with shales, conglomerates, and tuffs.

LGgw Gowlaun Member. *Conglomerate.*
Conglomerates and interbedded sandstones and mudstones.

LK Letterbrock Formation. *Greywacke.*
Mixed sequence of sandstones, conglomerates and green slates.

LM Lakes Marble Formation. *Mixed marble - schist.*
Mixed white to grey marbles, basic metavolcanics, schists and gritty psammitic schists and quartzites.

LN Lough Nacorra Formation. *Mixed schist - quartzite.*
Dominantly grey green psammitic schists, with pelites and quartzites.

LnMb Lough Nacorrussaun Metabasites. *Meta-basalt.*
Banded meta-basalts.

LT Lettermullen Formation. *Greywacke.*
Mixed sequence of conglomerates, sandstones, silt- and mudstones.

Md Metadolerite. *Meta-basalt.*
Metamorphosed dolerite intrusions, often schistose.

ME Mount Eagle Formation. *Quartzite.*
Pale white and grey quartzites and pebbly grits, minor schist.

Mg Metagabbro and Related Lithologies. *Meta-basalt.*
A group of metamorphosed, coarse grained intrusive bodies of basaltic and related types of rocks.

MK Lough Mask Formation. *Sandstone.*
Purple sandstones, with acidic lavas and breccias.

MKla Ardaun Lava Member. *Rhyolite.*
Mainly felsic ("granitic") lava, with interbedded mudstones.

MM Maam Formation. *Sandstone.*
Red, frequently pebbly sandstones and conglomerates with minor mudstones and siltstones.

MO Moy Sandstone Formation. *Sandstone.*
Grey sandstone and siltstone.

MP Mount Partry Formation. *Tuffs.*
Fine to coarse volcanic tuffs, breccias and sandstones and mudstones.

MT Maumtrasna Formation. *Conglomerate.*
Conglomerates and coarse sandstones.

MU Lough Muck Formation. *Sandstone.*
Sandstones and mudstones.

MW Mweelrea Formation. *Sandstone.*
Sandstones, with conglomerates locally, tuffs and slates.

MWsl Slate Members. *Slate.*
Green-grey slates.

NC North Carrowgarve Formation. *Schist.*
Pelitic and semi-pelitic schists, with some thin quartzites and psammites.

OD Ooghnadarve Formation. *Schist.*
Pelitic and semi-pelitic schists, locally graphitic and sometimes containing bright green serpentinite blocks. Mélange.

OmGr Omey Granite. *Granite.*
A named body of pink coloured granitic (Gr) rocks.

OuGr Oughterard Granite. *Granite.*
A medium to coarse grained body of non-porphyritic granite (Gr).

P Feldspar or Quartz Porphyry. *Granite.*
Fine grained, pale coloured granitic (Gr) rocks containing conspicuous crystals of feldspar or quartz.

Pg Paragneiss, Migmatite and Hornfels. *Meta-sediments.*
Mixed unit of highly metamorphosed sedimentary rocks.

PO Portnahally Formation. *Quartzite.*
Grey and white quartzites with thin semi-pelitic schists.

Qd Quartz Diorite Gneiss. *Meta-diorite.*
Highly metamorphosed and strongly banded rocks derived from an igneous rock intermediate in composition between basalt and granite.

Qg Quartz Diorite/Granite Gneiss. *Meta-diorite/Granite.*
As for Qd, but also including highly metamorphosed and banded granite.

RF Rockfleet Bay Limestone Formation. *Muddy limestone.*
Dark, fine limestone and calcareous shale.

RoGr Roundstone Granite. *Granite.*
A named body of a variety of coarse grained, pink - grey coloured granite.

RR Rosroe Formation. *Conglomerate.*
Coarse conglomerates with boulders up to 1m in a greenish sandstone matrix, with minor tuffs and black shales.

RY Ryans Farm Formation. *Meta-sandstone.*
Metamorphosed sandstones, mudstones and minor conglomerates and red and black cherts and lithic tuffs.

S Serpentinite. *Serpentinite.*
Metamorphosed and altered dark ultramafic igneous rock.

SA Salrock Formation. *Mudstones.*
Mainly mudstones and siltstones, generally red coloured, with tuffs near the base.

SC South Carrowgarve Formation. *Schist.*
Pelitic and semi-pelitic schists, locally with quartz pebbles and graphitic schists with basic volcanic clasts.

SCG SOUTH CONNEMARA GROUP. *Meta-volcanics and greywacke.*
Metamorphosed basic volcanic and intrusive rocks and greywacke sandstones and conglomerates.

SD Srahmore Lodge Dolomite Formation. *Marble.*
Brown dolomitic marble with minor quartzites and schists.

SH Sheeffry Formation. *Greywacke.*
Mainly interbedded grey-green greywacke, mudstones and thin tuff beds.

SI Skerd Formation. *Meta-basalt.*
Metamorphosed basic and acid volcanic and intrusive rocks, and slates.

SK Strake Banded Formation. *Siltstone.*
Red and grey siltstone, with minor sandstone and acid tuffs.

SR Sraheens Lough Formation. *Schist.*
Psammitic and subordinate semi-pelitic schists.

ST Streamstown Schist Formation. *Schist.*
Psammitic to semi-pelitic grey-brown schists with meta-basic igneous intrusions.

SV Skerdagh River Volcanic Formation. *Meta-volcanics.*
Basic metavolcanic lavas and tuffs.

TK Tourmakeady Formation. *Mixed limestone-tuff.*
Mixed unit of limestone, limestone breccias and fine to coarse volcanic tuffs and breccias.

tu Tuff Bands. *Tuff.*
Very fine volcanic ash deposits.

UD Undifferentiated Dalradian schists. *Schist.*
Undifferentiated schists.

WG Westport Grit Formation. *Meta-greywacke.*
Metamorphosed pebbly grits and greywackes and phyllites. Proximal turbidites.

APPENDIX 2: GLOSSARY OF TECHNICAL TERMS

ADIT: an horizontal tunnel or passage driven from the ground surface into a mine.

ALG(AE), ~AL: n., adj. a large group of almost exclusively aquatic plants, ranging from unicellular forms to giant kelps.

ALLUVIAL; ~ FAN: pertaining to deposition by a stream; fan-shaped mass of loose rock material where a stream issues from a mountain valley to a plain.

ALLUVIUM: sediments deposited by fluvial processes.

ANTICLINE: a fold, generally convex upwards, whose core contains the stratigraphically older rocks.

ARKOSE, ~IC: n., adj. a feldspar-rich sandstone.

AUREOLE: a zone surrounding an igneous intrusion in which the intruded rock is thermally metamorphosed.

BASALT: dark, fine-grained igneous rock, rich in iron and magnesium but little or no quartz.

BASIC: adj. applied to a largely quartz-free igneous rock, rocks low in silica.

BATHOLITH: large, generally cross-cutting pluton or group of plutons.

BED: (a) single unit of sedimentary rock, distinct from those on either side; (b) smallest formal lithostratigraphical unit.

BEDDING: arrangement of a sedimentary rock in layers of varying thickness and character.

BEDROCK: unweathered rock below the cover of soil, glacial deposits etc.

BLANKET BOG: a bog covering either flat or sloping ground, and depending on high rainfall, rather than accumulating in a local damp hollow.

BOULDER BED: a conglomerate containing boulder-size clasts, i.e. greater than 256mm across.

BOULDER CLAY: syn. till; an unsorted glacial deposit consisting of clay or rock flour with sub-angular stones of various sizes.

BRACHIOPOD: marine invertebrate of the phylum Brachiopoda (lamp-shell).

BRAIDED: branching and joining repeatedly to form an intricate network.

BRECCIA: coarse-grained clastic rock composed of angular broken rock fragments held together by mineral cement or fine-grained matrix.

CALCITE: a common rock forming mineral composed of calcium carbonate.

CALC-SILICATE: metamorphic rock, mainly of calcium-bearing silicates e.g. hornblende, diopside and wollastonite, formed by metamorphism of impure limestone or dolomite.

CARBONACEOUS: adj. describing (a) a rock rich in carbon; (b) a sediment containing organic matter, e.g. graphite, coal.

CARBONATE: a sediment formed by the organic or inorganic deposition from an aqueous solution of carbonates of calcium, magnesium or iron, e.g. limestone or dolomite.

CHERT: hard material composed of microcrystalline quartz or opaline silica, occurs as nodules or beds in limestone.

CLASTIC: adj. for rock or sediment composed mainly of broken fragments of pre-existing rocks/ minerals.

CLEAVAGE: finely spaced planar parting caused by compressive deformation of rocks (commonly forms oblique to bedding).

COMPLEX: a large-scale field association of different rocks having structural relations too complicated to be readily differentiated in mapping.

CONGLOMERATE: sedimentary rock comprising large rounded fragments (pebbles, cobbles to boulders) in a finer matrix.

CORAL: a marine invertebrate (class Anthozoa, phylum Coelenterata).

CORRIE: steep-walled hollow often on mountain side, commonly at head of a glacial valley.

CROSS STRATIFICATION: layers in a sedimentary or meta-sedimentary rock inclined at an angle to the general orientation of bedding. Generally reflect the depositional surfaces within wave or wind generated ripples or dunes.

CRUST : the outermost layer of the solid Earth. *continental* ~ - relatively low-density, compositionally evolved crust that forms the continents. *oceanic* ~ - basaltic crust formed at a spreading ridge and flooring the oceans.

DELTA: fan-shaped plain of alluvial sediments at river mouth crossed by many distributaries, often extending beyond the general trend of the coastline.

DESICCATION CRACK: irregular fracture (sometimes in a crudely polygonal pattern) formed by shrinkage of mud while drying under atmospheric conditions.

DIORITE: intermediate plutonic igneous rock composed of plagioclase (oligoclase, andesine), with amphibole (esp. hornblende), pyroxene, potassium feldspar up to 10%, and < 5% quartz.

DIP: the maximum angle of inclination of a bed of rock or planar fabric.

DOLERITE: syn. microgabbro or basalt; finer-grained variety of gabbro, intruded as thin sheets near the surface; *metadolerite* - metamorphosed dolerite.

DOLOMIT(E), ~IC: n., adj., calcium and magnesium-bearing carbonate mineral $CaMg(CO_3)_2$, also a rock composed of the mineral; containing the mineral dolomite.

DRIFT: (mining term) an horizontal or near horizontal underground working driven along the length of an ore zone, horizon or vein.

DRUMLIN: low smooth rounded elongate hill of glacial till.

DUCTILE: adj. pertaining to plastic deformation without fracturing.

DYKE: a sheet-like body of igneous rock cross-cutting other rock types.

ERRATIC: rock moved by glacial or floating ice from its original outcrop.

ESKER: long narrow sinuous ridge of sand and gravel deposited by a sub-glacial stream and left behind after the ice melted.

EXTRUSION, ~IVE: n., adj. igneous rock emplaced at the earth's surface; volcanic rock.

FAULT: planar fracture in rocks across which there has been some displacement; commonly inclined but may be vertical. *normal fault* - a fault in which the hanging wall has moved downward relative to the footwall. *reverse fault* - a fault in which the hanging wall has moved upward relative to the footwall. *wrench fault* - a fault in which the rocks on either side have moved laterally relative to each other.

FELDSPAR: white to pink mineral: a family of silicates of calcium, sodium, potassium and aluminium; see plagioclase ~.

FELSIC: acronym derived from *fe*ldspar + *l*enad (feldspathoid) + *si*lica + c, applied to igneous rock having abundant light coloured minerals.

FELSITE: informal name for a fine grained acid igneous rock.

FLAGGY: adj. applied to thinly-bedded sedimentary rock, usually parting easily (hence flagstones).

FLUVIATILE: syn. fluvial - of or pertaining to rivers.

FOLD: a curve or bend of a planar structure such as rock strata.

FORMATION: an association of related rocks which are capable of being mapped regionally.

FOSSIL: any remains or trace of an animal or plant preserved in rock.

GABBRO: coarsely crystalline intrusive igneous rock, iron and magnesium-rich with little or no quartz; *metagabbro* - metamorphosed gabbro.

GNEISS,~ OSE: n., adj. banded metamorphic rock formed at high temperatures and pressures.

GRANITE: pale, coarsely crystalline intrusive igneous rock, comprising mostly quartz and feldspar (potassium feldspar more than or equal to plagioclase).

GRANODIORITE: similar to granite, but with more plagioclase and mafic minerals.

GRAPHITIC: containing the mineral graphite - a soft black to steel grey mineral, native carbon.

GRAPTOLITE: fossil: extinct animal consisting of one or more fine branches with tiny cups (thecae).

GREYWACKE: dark-grey, impure, poorly sorted sandstone, composed of quartz, feldspar grains and rock fragments in a clayey matrix.

GROUNDMASS: the generally fine grained material surrounding coarser crystals in an igneous rock.

GROUP: ideally a succession of related formations, but sometimes applied to a thick but undivided rock unit.

HORNBLENDE: the commonest mineral of the amphibole group $(Ca,Na)_{2-3}(Mg,Fe^{+2},Al)_5(Al,Si)_8O_{22}(OH)_2$; commonly black, dark green, or brown.

HOT SPOT: a volcanic center, under oceans or on land, marking a persistent zone of rising hot material from the mantle of the Earth.

IGNEOUS: rock solidified from magma: may be extrusive or intrusive.

IGNIMBRITE: a rock or deposit of pumice, ash and rock fragments formed by a flow of volcanic ejecta.

INLIER: an area of rocks surrounded by rocks of younger age.

INTERGLACIAL: pertaining to a time interval between glacial stages.

INTERMEDIATE: adj. applied to igneous rock, transitional in composition between acid and basic, generally having silica content between 52-63 %.

INTRUS(ION), ~IVE: n., adj. igneous rock emplaced within the Earth's crust.

ISLAND-ARC: a chain of volcanic islands built on oceanic crust above a subduction zone.

KAME: mound, knob or ridge of stratified sand and gravel deposited by a sub-glacial stream, cf: **esker**.

KAME AND KETTLE TOPOGRAPHY: an undulating landscape in which a disordered assemblage of knolls, mounds or ridges of glacial drift with irregular depressions, pits or kettles that are commonly undrained and which may contain ponds or swamps.

KARST: topographic features formed by dissolution of rock, usually limestone, by surface water or groundwater, including sink holes, caves and underground drainage.

LAVA: general term for molten extrusive, or the rock solidified from it.

LIMB: the planar or gently curved part of a fold between more tightly curved hinge zones.

LIMESTONE: sedimentary rock composed of calcium carbonate, often containing fossils.

LITHIFICATION: processes whereby sediments are transformed into solid rocks.

LITHOLOGY: rock type.

LITHOSPHERE: the Earth's crust and uppermost part of the upper mantle that together make up the plates of the outer Earth; continent 100-150 km thick, ocean 70-80 km thick.

LITHOSTRATIGRAPHY: the organization of strata into units based on lithological character.

LITTORAL: pertaining to the marine environment between high and low water.

MAGMA: molten rock which will crystallize on cooling to form igneous rocks.

MAGMATIC ARC: a belt of magmatic activity above a subduction zone, including both volcanic and plutonic magmatism.

MANTLE: the zone below the Earth's crust and above the core.

MARBLE: metamorphosed limestone, formed by recrystallization at high temperature and pressure.

MASSIF: a large topographic or structural feature commonly formed of rocks more rigid or resistant than those of its surroundings.

MATRIX: fine-grained material filling the spaces between the larger grains in a sedimentary rock.

MÉLANGE: chaotic unit (sedimentary/tectono-sedimentary), deposited by mass-flow (large-scale slumping) of earlier unconsolidated sediments.

MEMBER: sub-unit of a formation, often mappable only locally.

META~: prefix to rock type etc. implying it is metamorphosed.

METAMORPH(ISM), ~IC, ~OSE: n., adj., vb. alteration of rocks by heat and/or pressure often accompanied by deformation; *~ic grade* - the intensity or rank of ~ism, "high", "medium" and "low" grade are used in a relative sense; *~ic facies* - a set of ~ic mineral assemblages associated with specific grade of ~ism.

MICA, ~ACEOUS: n., adj. sheet silicate mineral group, commonly containing potash, sodium, calcium, iron, magnesium, and aluminium; can be split perfectly into thin shiny flakes; most common types - muscovite (white), biotite (black/brown).

MICRO~: prefix to rock type etc. implying finer-grained variety.

MICROGABBRO: syn. dolerite.

MORAINE: mound or ridge of unsorted and unstratified glacial drift, mainly till, deposited commonly at margins and snout of glacier.

MUD,~STONE : fine-grained sediment/rock composed predominantly of clay.

MUSCOVITE: flaky, shiny colourless mica containing potassium and aluminium.

NORMAL FAULT: see under **Faults**.

OLD RED SANDSTONE: red beds of Devonian and basal Carboniferous age.

OLIVINE: iron and magnesium silicate mineral, $(Fe,Mg)_2(SiO)_4$.

OOLITE: type of limestone containing ooids.

OPHIOLIT(E), ~IC : n., adj. an association of mafic and ultramafic igneous rocks, and oceanic sediment, derived from oceanic lithosphere and structurally incorporated into continental crust.

OROGEN(Y), ~IC, ~ESIS: n., adj., n. mountain-forming event(s) or process: commonly a result of interactions involving convergence of lithospheric plates.

ORTHOGNEISS: gneiss formed by metamorphism of an igneous rock.

OVERTHRUST: a low-angle thrust fault of large scale, and with displacement generally measured in kilometres.

PEGMATITE: coarse/very coarse-grained igneous rock, representing final volatile-rich stage of magma crystallization; often forming sheet-like bodies.

PELIT(E),~ IC: n., adj. metamorphosed siltstone, mudstone or shale, commonly rich in micaceous minerals.

PERIDOTITE: an **ultramafic** igneous rock, characteristically rich in olivine. Frequently altered to **serpentinite**.

PILLOW LAVAS: lava that was extruded under water such that it solidified in rounded-"pillow-shaped" bodies.

PLAGIOCLASE, ~FELDSPAR: common rock-forming silicate with composition ranging from Albite (Ab) (sodium-rich - $NaAlSi_3O_8$) to Anorthite (An) (calcium-rich - $CaAl_2Si_2O_8$) extremes (An_0 - An_{100}).

PLATE, ~ TECTONICS: rigid thin portion of Earth's lithosphere; global theory of the movements and interactions of lithospheric plates.

PLAYA: a shallow desert basin, commonly coastal, usually with ephemeral lakes and evaporite deposits.

PLUTON, ~IC: n., adj. an igneous intrusion, commonly circular - elliptical; pertaining to rocks intruded at depth.

POTASSIUM FELDSPAR: (syn. K-feldspar) a pink mineral commonly: one of a family of silicates of calcium, sodium, potassium and aluminium characterised, in this instance, by notable amounts of potassium in the crystal structure.

PSAMMIT(E), ~IC: n., adj. metasedimentary rock comprising quartz with lesser amounts by volume of micas and feldspar.

PUMICE: a highly vesicular glassy volcanic rock, commonly rhyolitic in composition.

PYROXENE: group of commonly dark, rock-forming ferromagnesian silicate minerals.

PYROXENITE: an ultramafic plutonic rock chiefly composed of pyroxene.

QUARTZ: crystalline silica (SiO_2), very common hard rock-forming mineral.

QUARTZ DIORITE: a plutonic rock, having the composition of diorite but with an appreciable amount of quartz - the approximate intrusive equivalent of dacite.

QUARTZITE: a very hard sandstone composed mostly of quartz, where the quartz grains are completely and solidly cemented by secondary silica.

RADIOMETRIC AGE: age (in years) calculated from the quantitative determination of the decay of natural radioactive elements contained in the rock.

RAISE: an underground mine working driven upward from a lower to a higher level.

RED BEDS: sandstones, siltstones and shales, predominantly red due to iron oxide minerals; associated with arid/semi-arid environments.

REVERSE FAULT: see under **Faults**.

RHYOLITE: ~IC: n., adj. fine-grained igneous rock, rich in silica; the fine grained equivalent of granite.

RIPPLE-MARK: undulatory surface consisting of sub-parallel small-scale ridges and hollows formed at the interface between a fluid and incoherent sedimentary material; produced on land by wind and under water by currents or wave action.

ROCHE MOUTONNÉE: ice-sculpted rock form with elongate, smooth and domed surface, long-axis oriented in direction of ice-movement.

SANDSTONE: sedimentary rock, sand-sized grains, commonly of quartz.

SCHIST, ~OSE: n., adj. metamorphic rock with a schistosity: *schistosity* - coarse cleavage resulting from the growth of new platy minerals, particularly mica, during metamorphism accompanied by deformation.

SEDIMENTARY ROCK: lithified accumulation of clastic, organic, or chemically precipitated mineral grains.

SEMI-PELITE, ~IC: n., adj. metasedimentary rock comprising micas and quartz in about equal proportions and micas commonly > feldspar (syn. semi-pelitic schist).

SERICITE: a fine grained version of the micaceous mineral **muscovite**, commonly forming part of an alteration assemblage associated with mineral deposits.

SERPENTINITE: hydrated and mineralogically altered ultramafic igneous rock with serpentine group minerals, derived from e.g. olivine, pyroxene.

SHALE: mudstone with closely-spaced bedding-parallel partings.

SHARD: a glassy fragment in volcanic ash.

SHEAR: deformation resulting from stresses that cause parts of a body to slide relative to each other along their plane of contact.

SHEAR ZONE: a zone of any scale subjected to shearing stresses, may be ductile or brittle.

SILICA: silicon dioxide (SiO_2): quartz is the commonest of five forms.

SILICATE: common rock-forming minerals, divided into various groups of crystalline compounds whose structure contains SiO_4 tetrahedra; variously arranged as individual units, rings, chains, sheets, or networks.

SILL: igneous body with sheet-like form intruded parallel to bedding.

SILT: a sediment of particle size between 1/16 and 1/256mm.

SILTSTONE: sedimentary rock composed of silt-sized particles.

SLATE: metamorphosed mudstone with a near perfect closely-spaced, planar cleavage, hence can be split into slabs and thin plates.

SLIDE, TECTONIC~: ductile fault, commonly parallel with bedding, formed under pressure during orogenesis.

SMALL-SCALE FOLDING: folding seen in a hand specimen or a small outcrop.

STRATA: (singular stratum), syn. bed(s).

STRATIGRAPH(Y), ~ICAL: n., adj. (study of) all aspects of the science of rock strata; the sequence of rock strata.

STRIAE: ice-scratches on rock surfaces caused by debris carried by moving ice.

STRIKE: the direction of a horizontal line on an inclined bedding plane (or other planar fabric).

STRIKE-SLIP: in a fault, the component of movement that is parallel to the strike of the fault; a fault with displacement predominantly parallel to strike.

STROMATOLITE: a laminar algal deposit, which grows either subaerially or under water in warm climates; they are commonly preserved in shallow-water limestones.

SUBDUCTION, ~ZONE: the sinking of one lithospheric plate beneath the edge of another: ~ *zone* - the narrow linear zone of subduction, located beneath the oceanic trench, commonly dipping towards the continent.

SUCCESSION: stratigraphical sequence of rocks.

SUPERCONTINENT: an amalgamation of several continents.

SUPERGROUP: a formal assemblage of related groups, or of groups and formations.

SYNCLINE: fold which has the youngest rocks at its core, commonly concave-up.

TALC: a magnesium silicate mineral, very soft and with a soapy feel.

TECTONIC(S): said of forces involved in tectonics: study of regional scale structural or deformational features.

THRUST: shallowly inclined fault in which one unit has overridden another unit.

TILL, ~ITE: unsorted and unstratified, generally unconsolidated glacial drift; rock formed by lithification of till.

TRILOBITE: an extinct class of marine arthropod.

TUFF: a rock composed of compacted volcanic ash, crystals or rock fragments.

TURBIDIT(E), ~ITY CURRENT: sediment/rock deposited from *turbidity current*: density current of suspended sediment in water moving rapidly down a subaqueous (e.g. continental) slope.

ULTRAMAFIC: igneous/meta-igneous rock, composed principally of iron and magnesium rich silicate minerals, such as pyroxene and olivine.

UNCONFORMITY: an erosional break in the stratigraphical record at which strata truncate the structure of older strata.

VEIN: a planar or irregular, mineral filled fracture in any type of host rock. Usually composed of minerals conspicuously different from the host rock.

VESIC(LE), ~ULAR: n., adj. a spherical cavity in an igneous rock formed by solidification around a gas bubble: containing vesicles.

VOLCANIC: pertaining to the activity, structure, or rock types of a volcano.

VOLCANIC ARC: a line of volcanoes along a plate margin above a subduction zone.

WACKE: a "dirty" sandstone with poorly sorted angular mineral grains and rock fragments set in a fine grained clay/silt matrix.

THE GEOLOGICAL SURVEY OF IRELAND (GSI), founded in 1845, is Ireland's national earth science agency. The mandate of the GSI is to provide geological advice and information to a wide range of customers. Customers include Government, Local Authorities, industry, education, consultants and the general public. The main services which GSI provides are as follows:-

PUBLICATION SALES - A wide range of maps and publications on Irish geology are available.

ADVISORY SERVICES - are provided by the Bedrock, Quaternary/Geotechnical, Groundwater, Minerals and Marine Programmes.

GEOLOGICAL DATABASES - Each programme maintains databases which are being continuously developed.

DRILL CORE LIBRARY - GSI maintains an extensive range of drill core which is available for inspection by arrangement.

LIBRARY - A major source of geological information which is open to visitors 9.30 - 12.30 and 14.00 - 16.30, weekdays.

"DOWN TO EARTH" - A popular exhibition which provides an introduction to the nature and uses of geology and mineral resources. Open 14.00 - 16.30, weekdays, or by prior appointment. Advance booking for large groups is advisable.

PUBLIC OFFICE - Open 9.30 - 12.30 and 14.00 - 16.30, weekdays, for sales of maps and publications.

Geological Survey of Ireland,
Beggars Bush,
Haddington Road,
Dublin 4.

Tel: (01)6707444.
Fax: (01)6681782